Mathematics Series

Logarithms

A Selection of Classic Mathematical Articles Containing Examples and Exercises on the Subject of Algebra

By

Various Authors

Copyright © 2011 Read Books Ltd.
This book is copyright and may not be
reproduced or copied in any way without
the express permission of the publisher in writing

British Library Cataloguing-in-Publication Data
A catalogue record for this book is available from
the British Library

Contents

Elementary Algebra - Parts II and III.
A. W. Siddons and C. T. Daltry..*page* 1

Teach Yourself Algebra. P. Abbot..*page* 21

College Algebra. William H. Metzler
and Edward Drake Roe..*page* 41

A Treatise on Algebra, Embracing, Besides the Elementary
Principles, All the Higher Parts Usually Taught in Colleges.
George R. Perkins..*page* 68

Mathematics for the Practical Man. George Howe..*page* 95

A Treatise on Algebra. Oliver, Watt and Jones..*page* 106

Mathematics - Self-Taught - The Lubsen Method.
H. B. Lubsen..*page* 124

Elements of Algebra. Edward Atkins..*page* 155

An Academic Algebra. James M. Taylor..*page* 180

Modern Second Course in Algebra.
Webster Wells and Walter W. Hart..*page* 197

Elements of Algebra - Part I.
L. Carson and David Eugene Smith..*page* 209

INDICES. LOGARITHMS I

POSITIVE INDICES

26·1. Draw the graph of $y = 2^x$, where x is a positive integer, from $x = 1$ to $x = 4$. Take 1 inch as unit for x and $\frac{1}{2}$ inch as unit for y.

Note that the points obtained lie on a smooth curve. Can any meaning be attached to the intermediate points on the curve? In other words, can any meaning or value be attached to $2^{\frac{1}{2}}$ or $2^{1\cdot3}$?

26·2. Before answering this question we must consider our definition of a^n. On p. 133 we defined a^4 as $a \times a \times a \times a$; can we define $2^{\frac{1}{2}}$ or $2^{1\cdot3}$ in a similar way?

It appears that we have assigned no meaning to 2^x unless x is a positive integer. But let us draw the smooth curve, through the points we have determined, and see whether it suggests to us any interpretation of 2^x for other values of x.

From your graph read off the approximate values of $2^{1\cdot3}$, $2^{2\cdot7}$, $2^{3\cdot3}$.

26·3. Let us now look at the question from another point of view. On p. 137 we proved that

"In multiplying together powers of the same letter, add the indices; in dividing, subtract the index of the divisor from that of the dividend".

Or, in symbols, $a^p \times a^q = a^{p+q}$* and $a^p \div a^q = a^{p-q}$ (if $a \neq 0$) provided that p and q are positive integers and $p > q$ in the second case.

We also saw on p. 138 that $(a^p)^q = a^{pq}$.

Revise Ex. 11 c (i) and (ii), pp. 138, 139.

26·4. Now a^p and a^q have not yet been defined for values of p and q other than positive integers. Whatever meaning they may have when p and q are not positive integers, it would be convenient that the index laws should still hold.

We will now try to find meanings in special cases for $a^{\frac{x}{y}}$, a^0 assuming that the index laws, $a^p \times a^q = a^{p+q}$ and $a^p \div a^q = a^{p-q}$, hold for all rational† values of p and q.

---
* More strictly we should say $a^p \times a^q \equiv a^{p+q}$, but in future we shall frequently use = instead of ≡.

† That is for all values, positive or negative, which can be expressed as integers or as fractions with integral numerators or denominators.

ALGEBRA

According to our assumption
$$9^{\tfrac{1}{2}} \times 9^{\tfrac{1}{2}} = 9^{\tfrac{1}{2}+\tfrac{1}{2}} = 9^1 = 9.$$

Now what is that number which when multiplied by itself gives 9? What is the name given to the number? The ... of 9.*

Similarly $\quad 10^{\tfrac{2}{3}} \times 10^{\tfrac{2}{3}} \times 10^{\tfrac{2}{3}} = 10^{\tfrac{2}{3}+\tfrac{2}{3}+\tfrac{2}{3}} = 10^2.$

$$\therefore 10^{\tfrac{2}{3}} = \sqrt[3]{10^2}.$$

Again $\quad 10^0 \times 10^2 = 10^{0+2} = 10^2.$

Divide both sides by 10^2 and we have $10^0 = \dfrac{10^2}{10^2} = 1.$

26·5. This last result is so interesting that it is worthy of consideration from other view-points.

Note that $\quad 10^{\tfrac{1}{2}} = \sqrt{10} = 3 \cdot 162\ldots$ (see square root tables),
$$10^{\tfrac{1}{4}} = \sqrt{10^{\tfrac{1}{2}}} = \sqrt{3 \cdot 162} = 1 \cdot 779\ldots,$$
$$10^{\tfrac{1}{8}} = \sqrt{10^{\tfrac{1}{4}}} = \sqrt{1 \cdot 779} = 1 \cdot 333\ldots.$$

By continuing this process we can calculate 10^x for certain very small values of x; how does 10^x change as x gets smaller and **smaller**? Can we in this way ever get a number less than 1?

Again consider the table

10^3	10^2	10^1	10^0
1000	100	10	

What value does this suggest for 10^0?

26·6. We will now give a more general case.

By repeated application of our assumption
$$10^{\tfrac{p}{q}} \times 10^{\tfrac{p}{q}} \times 10^{\tfrac{p}{q}} \times \ldots \text{ to } q \text{ factors} \equiv 10^{\tfrac{p}{q}+\tfrac{p}{q}+\tfrac{p}{q}+\ldots \text{ to } q \text{ terms}},$$
$$\equiv 10^p.$$

Take the qth root of each side, $\therefore 10^{\tfrac{p}{q}} \equiv \sqrt[q]{10^p}.$

Note also that $10^{\tfrac{p}{q}} = \left(10^{\tfrac{1}{q}}\right)^p = (\sqrt[q]{10})^p.$

From the above work we see that 10^x has a meaning for the whole range of rational positive numbers from 0 upwards. Thus $10^{0 \cdot 3456} = 10^{\tfrac{3456}{10000}}$ and means $\sqrt[10000]{10^{3456}}.$

* In the present chapter we shall only concern ourselves with real positive roots.

INDICES. LOGARITHMS I

For the present all that we are concerned with is that such symbols as $10^{0.3456}$ and $10^{2.4972}$ have definite meanings; we are not much concerned with what those meanings are.

26·7. By aid of the square root tables (see § 26·5) and multiplication, find the values of

$$10^{\frac{1}{8}},\ 10^{\frac{1}{4}},\ 10^{\frac{3}{8}},\ 10^{\frac{1}{2}},\ 10^{\frac{5}{8}},\ 10^{\frac{3}{4}},\ 10^{\frac{7}{8}}.$$

Make a table for $y = 10^x$ thus:

x	0	0·125	0·25	0·375	0·5	
y	1				3·162	

Now draw the graph of $y = 10^x$ from $x = 0$ to $x = 1$, taking 5 in. as unit for x and 0·5 in. for y.*

Keep the graphs for Ex. 26 a, b. Fig. 26·1 shows the graph on a smaller scale.

EXERCISE 26 a

This exercise should be discussed in class. In each case use your graph $y = 10^x$ (or fig. 26·1).

1. Read off the values of $10^{0.1}$, $10^{0.15}$, $10^{0.2}$, $10^{0.25}$, etc.

2. Express the following as powers of 10: 1, 1·5, 2, 2·5, etc. [e.g. $6·5 = 10^{0.813}$].

3. From the graph $y = 10^x$ find the value of $10^{0.3}$, $10^{0.4}$. By multiplication find the value of $10^{0.3} \times 10^{0.4}$ and compare the result with $10^{0.3+0.4}$, i.e. $10^{0.7}$.

4. Find a and b in the following cases: $2 = 10^a$, $3 = 10^b$. Now $2 \times 3 = 10^a \times 10^b = 10^{a+b}$. From the graph find the value of 10^{a+b}; is this what you expect?

5. Copy the following and fill up the blanks, using the graph where necessary:

$$2·65 \times 1·7 = 10^{\cdots} \times 10^{\cdots} = 10^{\cdots + \cdots} = 10^{\cdots} = \ldots.$$

Check by multiplication.

6. Use the graph to perform the following multiplications:

(i) $3·75 \times 2$, (ii) $1·45 \times 5·25 \times 1·15$, (iii) $2·5^2$.

* If large enough paper is available, easier scales would be 10 in. as unit for x and 1 in. for y.

7. Find a and b in the following cases: $9 = 10^a$, $2 = 10^b$. Now $9 \div 2 = 10^a \div 10^b = 10^{a-b}$. From the graph find the value of 10^{a-b}; is the value correct?

8. Copy the following and fill up the blanks:
$$2{\cdot}65 \div 1{\cdot}7 = 10^{\cdots} \div 10^{\cdots} = 10^{\cdots - \cdots} = 10^{\cdots} = \ldots$$
Check by division.

9. Use the graph to simplify the following:

(i) $9{\cdot}75 \div 4$,   (ii) $5{\cdot}25 \div 4{\cdot}85$,   (iii) $3{\cdot}75 \times 5{\cdot}9 \div 2{\cdot}7$.

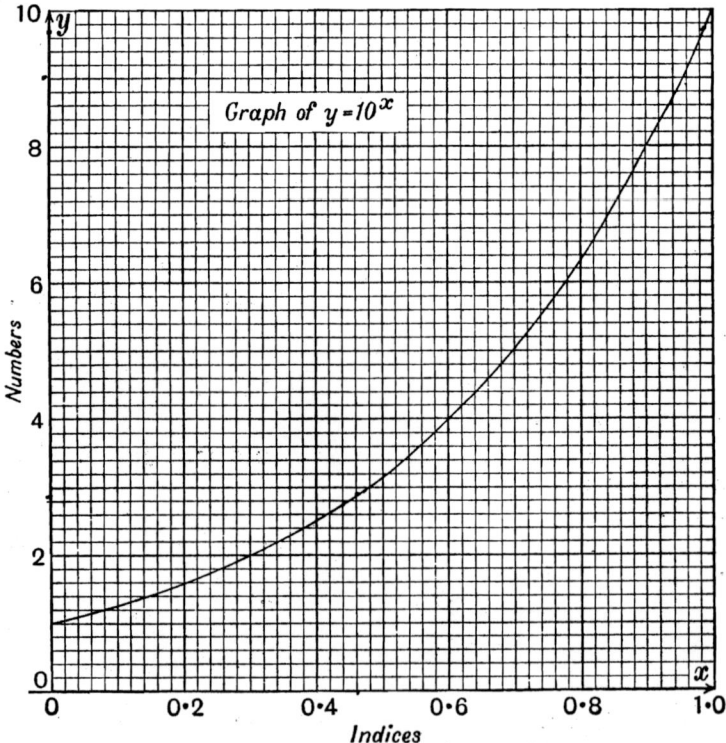

Fig. 26·1

INDICES. LOGARITHMS I

26·8. Note that, if n is a number between 1 and 10, there is a corresponding index x between 0 and 1 which makes $n = 10^x$ and *vice versa*.

We shall speak of a number between 1 and 10 (e.g. 3·142, 7·04, 5) as a number in standard form.

Now $2 \cdot 16 = 10^{0 \cdot 335}$.

$$\therefore 2\!\downarrow\!1 \cdot 6^* = 2 \cdot 16 \times 10 = 10^{0 \cdot 335} \times 10^1 = 10^{1 \cdot 335}.$$

Again $\quad 2\!\downarrow\!16 = 2 \cdot 16 \times 10^2 = 10^{0 \cdot 335} \times 10^2 = 10^{2 \cdot 335}.$

EXERCISE 26 b

Throughout this exercise use arrowheads as suggested in the footnote.

1. Copy the following and fill up the gaps:

$2\!\downarrow\!1 \cdot 6 = 10^1 \times 2 \cdot 16 = 10^{1 \cdot 335}$, $3\!\downarrow\!1 \cdot 4 = 10^{\cdots} \times 3 \cdot 14 = 10^{\cdots}$,

$2\!\downarrow\!160 = 10^{\cdots} \times 2 \cdot 16 = 10^{\cdots}$, $3\!\downarrow\!14 = 10^{\cdots} \times 3 \cdot 14 = 10^{\cdots}$,

$2\!\downarrow\!1600 = 10^{\cdots} \times 2 \cdot 16 = 10^{\cdots}$, $8\!\downarrow\!270 = 10^{\cdots} \times 8 \cdot 27 = 10^{\cdots}$.

2. If the following numbers are expressed as powers of 10, what is the integral part of each index?

(i) 276,   (ii) 3·14,   (iii) 1728,   (iv) 1933,
(v) 93,000,000,   (vi) 15,   (vii) 10,000.

3. $3 \cdot 22 = 10^{0 \cdot 508}$, $6 \cdot 78 = 10^{0 \cdot 831}$. Use these facts to express the following in the form 10^x:

(i) 322,   (ii) 67·8,   (iii) 32200,   (iv) 678000.

4. $10^{0 \cdot 6435} = 4 \cdot 4$ and $10^{0 \cdot 7443} = 5 \cdot 55$. Use these facts to find the values of

(i) $10^{2 \cdot 6435}$,   (ii) $10^{3 \cdot 7443}$,   (iii) $10^{1 \cdot 6435}$,
(iv) $10^{5 \cdot 7443}$,   (v) $10^{2 \cdot 7443}$,   (vi) $10^{4 \cdot 6435}$.

Use indices to find approximately the values of the following:

5. $276 \times 3 \cdot 14$.   **6.** $37 \cdot 3 \times 25 \cdot 5$.   **7.** $276 \div 3 \cdot 14$.
8. $37 \cdot 3 \div 25 \cdot 5$.   **9.** $(27)^2 \times 555$.   **10.** $255 \div 37 \cdot 3$.

* It will be found useful for the present to put in an arrowhead to indicate the place which the decimal point would occupy if the number were changed into standard form.

ALGEBRA [26·9

26·9. If $n = 10^x$, we say that x is the index corresponding to n, the **base** being 10; or we say x is the **logarithm** of n for the base 10.

We choose 10 for base because ordinary numbers are expressed in powers of 10 (256 means $2 \times 10^2 + 5 \times 10^1 + 6 \times 10^0$). This enables the index (or logarithm) for any number to be found when once we know the indices (or logarithms) for numbers between 1 and 10.

USE OF TABLES

26·10. So far we have used a graph to determine the indices corresponding to given numbers. By drawing larger and more accurate graphs it would be possible to determine the indices to more significant figures, but it is easier to use tables.

Look at the following, which is part of the table* given at the end of the book.

	0	1	2	3	4	5	6	7	8	9
55	·7404	·7412	·7419	·7427	·7435	·7443	·7451	·7459	·7466	·7474
56	·7482	·7490	·7497	·7505	·7513	·7520	·7528	·7536	·7543	·7551
57	·7559	·7566	·7574	·7582	·7589	·7597	·7604	·7612	·7619	·7627
58	·7634	·7642	·7649	·7657	·7664	·7672	·7679	·7686	·7694	·7701
59	·7709	·7716	·7723	·7731	·7738	·7745	·7752	·7760	·7767	·7774

The figures in heavier type are numbers whose corresponding indices are given in the lighter type to the right, the base being 10.

Notice the following points:

In finding the index corresponding to a given number

   (i) **The significant figures in the number determine the fractional part† of the index and have nothing to do with the integral part.‡**

   (ii) **The position of the decimal point in the number determines the integral part of the index and does not affect the fractional part.**

If the number is in standard form (i.e. has one figure to the left of the decimal point), the integral part of the index is 0, and the 0 should be written down.

---

* In the table at the end of the book, and in most logarithm tables, the decimal point is omitted throughout; it is then understood that the point comes to the left of each of the given indices.

† Often called the **mantissa**.   ‡ Often called the **characteristic**.

INDICES. LOGARITHMS I

If the number is not in standard form, consider by what power of 10 the standard form must be multiplied to obtain the given number.

In finding the number corresponding to a given index

(iii) **The fractional part of the index determines the significant figures in the number; begin by considering the fractional part.**

(iv) **The integral part of the index determines the position of the decimal point in the number.**

If the integral part of the index is 0, the number is in standard form.

If the integral part of the index is not 0, consider how many places the decimal point must be moved from the standard position.

EXERCISE 26 c (Oral)

1. From the tables at the end of the book read off the indices corresponding to

(i) 3·14, (ii) 1·41, (iii) 6·7,
(iv) 9,   (v) 8·88, (vi) 1·03.

2. From the tables read off to three significant figures the numbers corresponding to the following indices; in case the exact index is not given in the tables take the nearest index:

(i) 0·6628, (ii) 0·6990, (iii) 0·7284,
(iv) 0·8471, (v) 0·9140, (vi) 0·9820.

3. Read off indices corresponding to

(i) 6·789, (ii) 9·876, (iii) 1·753, (iv) 2·056.

4. Read off, to four significant figures, the numbers corresponding to the following indices:

(i) 0·4396, (ii) 0·4878, (iii) 0·6617,
(iv) 0·6932, (v) 0·4261, (vi) 0·2689.

5. Read off the indices corresponding to

(i) 27·27, (ii) 451·5, (iii) 1728,
(iv) 625,   (v) 60000,   (vi) 75·79.

6. Read off, to four significant figures, the numbers corresponding to the following indices:

(i) 2·8654, (ii) 4·9515, (iii) 1·9562,
(iv) 3·9031, (v) 2·4419, (vi) 5·5555.

MULTIPLICATION AND DIVISION

26·11. In using indices, or logarithms as they are generally called, for multiplication or division it is convenient to arrange the work as follows:*

Example 1. To find the value of $3 \cdot 142 \times 25 \cdot 37 \times 457 \cdot 6$.

$$3 \cdot 142 \times 25 \cdot 37 \times 457 \cdot 6 = 10^{0 \cdot 4972} \times 10^{1 \cdot 4043} \times 10^{2 \cdot 6605}$$
$$= 10^{4 \cdot 5620}$$
$$= 36470.$$

The reader is advised to use powers of 10, as above, till he is thoroughly familiar with the use of logarithms. Afterwards, but not too soon, he should adopt the following arrangement:

$3 \cdot 142 \times 25 \cdot 37 \times 457 \cdot 6 = 36470.$

No.	Log
3·142	0·4972
25·37	1·4043
457·6	2·6605
Product	4·5620

Example 2. To find the value of

$$73 \cdot 35 \times \frac{273}{273 + 21 \cdot 3} \times \frac{760 - 18 \cdot 5}{760}.$$

Expression

$$= 73 \cdot 35 \times \frac{273}{294 \cdot 3} \times \frac{741 \cdot 5}{760}$$

$$= 66 \cdot 40.$$

No.	Log	
73·35	1·8654	
273	2·4362	
741·5	2·8701	
Numerator		7·1717
294·3	2·4687	
760	2·8808	
Denominator		5·3495
Fraction		1·8222

* It should be noted that when 4-figure tables are used the fourth significant figure is liable to error.

INDICES. LOGARITHMS I

EXERCISE 26 d (i)
Find the values of the following:
1. $3\cdot142 \times 88\cdot86$.
2. $283\cdot7 \times 5943$.
3. $256\cdot5 \div 43\cdot47$.
4. $6562 \div 3\cdot142$.
5. $143\cdot7 \times 12\cdot05$.
6. $63\cdot28 \div 25\cdot52$.
7. $(273\cdot3)^2$.
8. $(8\cdot517)^3$.
9. $(4020)^3$.
10. $(3\cdot142)^2$.
11. $16\cdot1 \times (2\cdot72)^2$.
12. $3\cdot142 \times (1\cdot315)^2$.
13. $\dfrac{15\cdot73}{8\cdot621}$.
14. $\dfrac{8\cdot754 \times 9\cdot361}{12\cdot07}$.
15. $\dfrac{756\cdot3 \times 40\cdot04}{1500}$.
16. $\dfrac{474700 \times 21\cdot25}{6774}$.
17. $\dfrac{9}{2\cdot718 \times 1\cdot552}$.
18. $\dfrac{371\cdot2 \times 22\cdot43}{300\cdot7}$.

EXERCISE 26 d (ii)
Find the values of the following:
1. $1\cdot406 \times 426\cdot1$.
2. $20\cdot1 \times 196\cdot7$.
3. $30\cdot91 \div 10\cdot4$.
4. $19\cdot88 \div 1\cdot497$.
5. $37\cdot13 \times 54\cdot41$.
6. $8123 \div 907$.
7. $(92\cdot41)^2$.
8. $(5\cdot004)^3$.
9. $(203)^3$.
10. $(17\cdot89)^2$.
11. $32\cdot23 \times (15\cdot75)^2$.
12. $3\cdot142 \times (25\cdot5)^2$.
13. $\dfrac{8979}{5678}$.
14. $\dfrac{1052 \times 2\cdot171}{990\cdot3}$.
15. $\dfrac{37\cdot13 \times 3\cdot713}{24}$.
16. $\dfrac{2240 \times 17\cdot18}{124\cdot4}$.
17. $\dfrac{2}{1\cdot239 \times 1\cdot567}$.
18. $\dfrac{77\cdot07 \times 1357}{98760}$.

POWERS AND ROOTS

26·12. In some of the foregoing examples we have used indices or logarithms to find the squares and cubes of given numbers.

We see at once that if $n = 10^q$, then
$$n^2 = n \times n = 10^q \times 10^q = 10^{q+q} = 10^{2q}.$$
Similarly $\quad n^3 = 10^q \times 10^q \times 10^q = 10^{q+q+q} = 10^{3q}.$

More generally if p is any whole number
$$n^p = (10^q)^p = 10^{pq} \quad \text{(see p. 138)}.$$
Hence we can find a power of a given number.

26·13. Again, suppose that we want to find the square root of a given number n; let 10^y be the square root.

Then $\qquad 10^y \times 10^y = n,$

$\therefore 10^{2y} = n = 10^q$ suppose.

$\therefore 2y = q,$

$\therefore y = \dfrac{q}{2}.$

$\therefore$ the logarithm of $\sqrt{n} = \frac{1}{2}$ of the logarithm of n.

Similarly the logarithm of $\sqrt[3]{n} = \frac{1}{3}$ of the logarithm of n.

Exercise 26 e (i)

Use logarithms to evaluate the following:

1. $\sqrt{323\cdot7}.$
2. $\sqrt{32\cdot37}.$
3. $\sqrt{3\cdot237}.$
4. $\sqrt[3]{5275}.$
5. $\sqrt[3]{527\cdot5}.$
6. $\sqrt[3]{52\cdot75}.$
7. $\sqrt[3]{5\cdot275}.$
8. $\sqrt{3}.$
9. $\sqrt{2}.$
10. $\sqrt{20}.$
11. $\sqrt{10}.$
12. $\sqrt{\dfrac{570\cdot2}{3\cdot142}}.$
13. $\sqrt[3]{\dfrac{250}{4\cdot771}}.$
14. $\sqrt{\dfrac{73\cdot2}{2\cdot563 \times 4\cdot4}}.$
15. $\sqrt{\dfrac{144 \times 25\cdot6}{736\cdot2}}.$
16. $\sqrt[3]{\dfrac{3 \times 762\cdot8}{4 \times 3\cdot142}}.$
17. $\sqrt[3]{\dfrac{20\cdot07 \times 51\cdot37}{14\cdot72 \times 2\cdot854}}.$
18. $2 \times 3\cdot142 \sqrt{\dfrac{35}{32\cdot2}}.$

Exercise 26 e (ii)

Use logarithms to evaluate the following:

1. $\sqrt{456\cdot6}.$
2. $\sqrt{45\cdot66}.$
3. $\sqrt{4\cdot566}.$
4. $\sqrt[3]{6786}.$
5. $\sqrt[3]{678\cdot6}.$
6. $\sqrt[3]{67\cdot86}.$
7. $\sqrt[3]{6\cdot786}.$
8. $\sqrt{5}.$
9. $\sqrt{8}.$

INDICES. LOGARITHMS I

10. $\sqrt{70}$.   11. $\sqrt[3]{100}$.   12. $\sqrt{\dfrac{299}{3\cdot 142}}$.

13. $\sqrt[3]{\dfrac{3000}{2\cdot 456}}$.   14. $\sqrt{\dfrac{4}{2\cdot 012 \times 1\cdot 39}}$.   15. $\sqrt{\dfrac{981 \times 350}{2\cdot 225}}$.

16. $\sqrt[3]{\dfrac{1728 \times 3}{4 \times 3\cdot 142}}$.   17. $\sqrt[3]{\dfrac{98\cdot 77 \times 2\cdot 452}{10\cdot 79 \times 20\cdot 09}}$.

18. $2 \times 3\cdot 142 \sqrt{\dfrac{73}{32\cdot 2}}$.

EXERCISE 26 f

Take $\pi = 3\cdot 142$.

1. $s = \tfrac{1}{2}gt^2$, find s when (i) $g = 32\cdot 2$, $t = 20\cdot 5$; (ii) $g = 981$, $t = 16\cdot 5$.

2. $M = pr^n$, find M when (i) $p = 80$, $r = 1\cdot 025$, $n = 2$; (ii) $p = 93\cdot 75$, $r = 1\cdot 03$, $n = 4$.

3. $A = \pi r^2$, find A when (i) $r = 5\cdot 5$; (ii) $r = 4\cdot 25$.

4. $S = 4\pi r^2$, find S when (i) $r = 3\cdot 17$; (ii) $r = 20\cdot 8$.

5. $V = \tfrac{4}{3}\pi r^3$, find V when (i) $r = 5\cdot 875$; (ii) $r = 25\cdot 7$.

6. $V = \tfrac{1}{3}\pi r^2 h$, find V when (i) $r = 2\cdot 75$, $h = 4\cdot 5$; (ii) $r = 3\cdot 25$, $h = 7\cdot 58$.

7. $t = 2\pi\sqrt{\dfrac{l}{g}}$, find t when (i) $l = 42$, $g = 32\cdot 2$; (ii) $l = 1500$, $g = 981$.

8. $A = \tfrac{1}{4}a^2\sqrt{3}$, find A when (i) $a = 6\cdot 6$; (ii) $a = 22\cdot 72$.

9. $D = \sqrt[3]{\dfrac{65H}{N}}$, find D when (i) $H = 130$, $N = 125$; (ii) $H = 150$, $N = 155$.

10. $d = \sqrt{\dfrac{2rh}{5280}}$, find d when (i) $r = 3950$, $h = 117$; (ii) $r = 3950$, $h = 490$.

11. Find the number of digits in (i) 3^{20}; (ii) 2^{64}.

12. Evaluate (i) $1\cdot 05^{10} - 1$; (ii) $2\cdot 05^5 - 2^5$.

ALGEBRA Ex. 26*f*

13. Find the value of

(i) $70.4 \times \dfrac{764.3 - 14.6}{760} \times \dfrac{273}{273 + 17.2}$;

(ii) $\dfrac{2.873 \times 35.97 \times 273500}{59.7 + 12.03 \times 80.5}$.

14. $A = P\left(1 + \dfrac{r}{100}\right)^n$, find A when

(i) $P = 2000$, $r = 3\frac{1}{2}$, $n = 4$;

(ii) $P = 2500$, $r = 2\frac{1}{2}$, $n = 3$.

15. $V = \dfrac{(t + 461)v}{T + 461}$, find V when

(i) $t = 700$, $T = 1100$, $v = 1050$;

(ii) $t = 800$, $T = 1200$, $v = 1000$.

16. $f = \dfrac{w + 10}{2240W} \times \dfrac{V^2}{2S}$, find f when

(i) $w = 120$, $W = 7.5$, $V = 2400$, $S = 19.8$;

(ii) $w = 100$, $W = 6.5$, $V = 2000$, $S = 17.4$.

17. $H = 49000 \left(\dfrac{R - r}{R + r}\right)\left(1 + \dfrac{T + t}{900}\right)$, find H when

(i) $R = 29.60$, $r = 25.35$, $T = 67$, $t = 32$;

(ii) $R = 28.75$, $r = 26.75$, $T = 68$, $t = 29$.

18. $V = \pi(R^2 - r^2)l$, find V when

(i) $R = 74.35$, $r = 42.63$, $l = 132.8$;

(ii) $R = 17.34$, $r = 13.68$, $l = 125.4$.

19. $A = \sqrt{s(s-a)(s-b)(s-c)}$, where $s = \frac{1}{2}(a + b + c)$, find A when

(i) $a = 32.8$, $b = 43.6$, $c = 70.1$;

(ii) $a = 13.5$, $b = 14.8$, $c = 17.6$.

20. $C = \sqrt{\dfrac{s(s-c)}{(s-a)(s-b)}}$, where $s = \frac{1}{2}(a + b + c)$, find C when

(i) $a = 3.56$, $b = 4.75$, $c = 5.33$;

(ii) $a = 120.5$, $b = 127.4$, $c = 91.7$.

INDICES. LOGARITHMS I

NEGATIVE INDICES

26·14. So far we have dealt only with positive indices; we must now consider whether a negative index can have a meaning.

We shall proceed as in the case of fractional indices, and make the assumption that the laws $10^m \times 10^n \equiv 10^{m+n}$ and $10^m \div 10^n \equiv 10^{m-n}$ are true also when m and n are negative numbers.

Consider the table on the right.

In the left hand column 1 is subtracted from each index to get the one below it; what is done in the right hand column? What should be inserted in the blanks?

10^2	100
10^1	10
10^0	1
10^{-1}	
10^{-2}	

Again, what does $10^m \div 10^n \equiv 10^{m-n}$ become when $m = 3$ and $n = 5$?

We have $$10^3 \div 10^5 = 10^{-2},$$
$$\therefore 10^{-2} = \frac{1}{10^2}.$$

Again, consider 10^{-7},
$$10^{-7} \times 10^7 = 10^{-7+7} = 10^0 = 1 \quad \text{(see § 26·4),}$$

$\therefore$ dividing both sides by 10^7, $10^{-7} = \dfrac{1}{10^7}$.

Similarly $$10^{-\frac{1}{2}} = \frac{1}{10^{\frac{1}{2}}}.$$

26·15. We find from the tables
$$1\cdot 224 = 10^{0\cdot 0878}.$$
Now
$$\cdot 01224 = 1\cdot 224 \div 100$$
$$= 10^{0\cdot 0878} \div 10^2$$
$$= 10^{0\cdot 0878 - 2}.$$

This index is of course equal to $-1\cdot 9122$ (which is $-1 - 0\cdot 9122$), but it is more convenient to express it otherwise. It is generally written thus $\bar{2}\cdot 0878$, and this must be taken to mean $-2 + 0\cdot 0878$.

It is important always to keep the fractional part of the index positive, for all the indices given in the tables are *positive* fractions; thus

$$10^{-1 \cdot 9122} = \frac{1}{10^{1 \cdot 9122}} = \frac{1}{81 \cdot 70} \quad \text{(see § 26·14)}.$$

Now $\quad 10^{-1 \cdot 9122} = 10^{\bar{2} \cdot 0878} = \cdot 01224.$

The latter is obviously the more useful result.

EXERCISE 26 g

1. On squared paper draw a pair of axes, take 1 inch as unit and graduate each axis $\bar{3}, \bar{2}, \bar{1}, 0, 1, 2, 3$. Mark the following points and write the coordinates against each point: A, (0, 2·4); B, (0, $\bar{3}$·2); C, (1, 1·4); D, (1, $\bar{1}$·4); E, (1·5, $\bar{2}$·7); F, ($\bar{1}$, 0·5); G, (2, $\bar{3}$·7).

2. Copy the following table and fill up the blanks as in the first line:

0·03142 ↓ 0·7777 ↓ 0·0054	$10^{-2} \times 3 \cdot 142$	$10^{\bar{2} \cdot 4972}$	500 ↓ 0·0006 ↓ 0·3754		

3. If the following numbers are expressed as powers of 10, what is the integral part of each index?

(i) 0·04, (ii) 365, (iii) 0·00057, (iv) 0·3333, (v) 52, (vi) 0·00728, (vii) 7·35.

4. $10^{0 \cdot 3010} = 2$ and $10^{0 \cdot 4771} = 3$. Use these facts to find the values of

(i) $10^{2 \cdot 3010}$, (ii) $10^{\bar{2} \cdot 3010}$, (iii) $10^{\bar{1} \cdot 4771}$,

(iv) $10^{1 \cdot 4771}$, (v) $10^{\bar{4} \cdot 3010}$, (vi) $10^{4 \cdot 3010}$.

26·16. In dealing with logarithms such as $\bar{1} \cdot 3476$, it is important to remember that

(i) $\bar{1} \cdot 3476$ stands for $-1 + 0 \cdot 3476$;

(ii) the fractional part of the logarithm should always be kept positive.

INDICES. LOGARITHMS I

At first the simplest plan is to write out the integral part in full with the proper signs attached (this should be done in the margin). Thus

$\bar{2}\cdot 7 + 3\cdot 8 = 2\cdot 5$ $\qquad$ $1^{*} + (-2) + 3 = 1 - 2 + 3 = 2,$

$\bar{3}\cdot 7 + 2\cdot 4 + \bar{1}\cdot 2 = \bar{1}\cdot 3$ $\qquad$ $1^{*} - 3 + 2 - 1 = -1,$

$\bar{3}\cdot 2 - \bar{2}\cdot 7 = \bar{2}\cdot 5$ $\qquad$ $-1^{*} - 3 - (-2) = -2.$

Multiplication is done in the same way:

$\bar{1}\cdot 4 \times 3 = 1\cdot 2^{*} + 3(-1) = \bar{2}\cdot 2 \qquad 1 - 3 = -2.$

In the case of division it is necessary to make a negative integer up to one that is exactly divisible by the divisor, thus:

$$\bar{3}\cdot 6 \div 2 = (-4 + 1\cdot 6) \div 2 = -2 + \cdot 8 = \bar{2}\cdot 8,$$
$$\bar{4}\cdot 7 \div 3 = (-6 + 2\cdot 7) \div 3 = \bar{2}\cdot 9.$$

EXERCISE 26 h (Oral)†

In every case make the fractional part of the result positive.

1. Add

	(i)	(ii)	(iii)	(iv)	(v)
	$3\cdot 3$	$\bar{3}\cdot 3$	$2\cdot 7$	$\bar{2}\cdot 7$	$\bar{2}\cdot 7$
	$\bar{2}\cdot 4$	$2\cdot 4$	$\bar{3}\cdot 3$	$3\cdot 3$	$\bar{3}\cdot 3$
	$1\cdot 2$	$\bar{1}\cdot 2$	$0\cdot 4$	$1\cdot 4$	$1\cdot 4$

2. Subtract the lower number from the upper

(i) $3\cdot 5$ / $\bar{1}\cdot 3$ $\quad$ (ii) $\bar{4}\cdot 7$ / $2\cdot 3$ $\quad$ (iii) $\bar{4}\cdot 7$ / $\bar{2}\cdot 2$ $\quad$ (iv) $\bar{2}\cdot 3$ / $3\cdot 7$ $\quad$ (v) $\bar{2}\cdot 3$ / $\bar{3}\cdot 8$

(vi) $1\cdot 3$ / $2\cdot 1$ $\quad$ (vii) $1\cdot 3$ / $3\cdot 5$ $\quad$ (viii) $3\cdot 1$ / $\bar{1}\cdot 3$ $\quad$ (ix) $0\cdot 0$ / $1\cdot 6$ $\quad$ (x) $0\cdot 0$ / $\bar{1}\cdot 4$

3. Simplify

(i) $1\cdot 4 \times 3$, $\quad$ (ii) $\bar{1}\cdot 2 \times 3$, $\quad$ (iii) $\bar{1}\cdot 4 \times 3$, $\quad$ (iv) $\bar{2}\cdot 5 \times 3$,

(v) $2\cdot 8 \div 2$, $\quad$ (vi) $\bar{2}\cdot 8 \div 2$, $\quad$ (vii) $\bar{2}\cdot 8 \div 3$, $\quad$ (viii) $\bar{2}\cdot 8 \div 4$,

(ix) $\bar{8}\cdot 8 \div 3$, $\quad$ (x) $\bar{7}\cdot 8 \div 2$, $\quad$ (xi) $\bar{7}\cdot 6 \div 4$, $\quad$ (xii) $\bar{3}\cdot 9 \div 10$.

* From adding, subtracting or multiplying the fractional parts.
† If further examples are needed, it is easy to make them up.

ALGEBRA [26·16

EXERCISE 26 i (i)

Evaluate

1. 0.3594×25.82.
2. 0.0587×2.078.
3. 0.7836×4.774.
4. $0.8532 \div 9.761$.
5. $0.08532 \div 97.61$.
6. $976.1 \div 0.08532$.
7. $(0.0227)^2$.
8. $(0.6015)^2$.
9. $(0.03714)^3$.
10. $(0.3437)^3$.
11. $\sqrt{0.03571}$.
12. $\sqrt{0.3571}$.
13. $\sqrt[3]{0.3571}$.
14. $\sqrt[3]{0.03571}$.
15. $\dfrac{1}{72.64}$.
16. $\dfrac{1}{3.142}$.
17. $\dfrac{1}{0.7071}$.
18. $\dfrac{1}{\sqrt{22.75}}$.
19. $\sqrt[3]{\dfrac{1}{85.77}}$.
20. $\sqrt{\dfrac{1}{0.2544}}$.
21. $\sqrt[3]{\dfrac{1}{0.0717}}$.
22. $\sqrt[3]{\dfrac{1}{3.759^2}}$.
23. $0.5712 \times 763.28 \times 0.0715$.
24. $42.73 \times 0.07562 \times 0.0086$.
25. $3.142 \times 37.62 \div 0.2335$.
26. $1.097 \times 0.0823 \div 0.9036$.
27. $\dfrac{0.7361 \times 0.03715}{2.165 \times 0.8717}$.
28. $\dfrac{53.84 \times 70.34}{827.1 \times 256.2}$.
29. $\dfrac{326.5 \times 41.96}{21.6 \times 0.0357}$.
30. $\sqrt{\dfrac{666.6 \times 0.7317}{0.0861 \times 2.654}}$.
31. $\sqrt[3]{\dfrac{0.07624}{3.142 \times 27.05}}$.
32. $\sqrt{\dfrac{209.3 \times 0.0168}{(16.91)^3}}$.

EXERCISE 26 i (ii)

Evaluate

1. 0.1934×439.1.
2. 0.0952×2500.
3. 0.7861×7.861.
4. $1.237 \div 83.72$.
5. $0.1237 \div 8.372$.
6. $837.2 \div 0.01237$.
7. $(0.09721)^2$.
8. $(0.209)^2$.
9. $(0.07624)^3$.
10. $(0.1416)^3$.
11. $\sqrt{0.05986}$.
12. $\sqrt{0.5986}$.
13. $\sqrt[3]{0.5986}$.
14. $\sqrt[3]{0.05986}$.
15. $\dfrac{1}{273}$.
16. $\dfrac{1}{1.732}$.
17. $\dfrac{1}{0.6261}$.
18. $\dfrac{1}{\sqrt{47.86}}$.
19. $\sqrt[3]{\dfrac{1}{72.84}}$.
20. $\sqrt{\dfrac{1}{0.1934}}$.
21. $\sqrt[3]{\dfrac{1}{0.0876}}$.
22. $\sqrt[3]{\dfrac{1}{2.894^2}}$.

INDICES. LOGARITHMS I

23. $1728 \times 0{\cdot}155 \times 0{\cdot}07934.$ **24.** $37{\cdot}26 \times 0{\cdot}4006 \div 0{\cdot}0029.$

25. $3700 \times 0{\cdot}00593 \div 230{\cdot}9.$ **26.** $110{\cdot}8 \times 0{\cdot}003719 \div 0{\cdot}1842.$

27. $\dfrac{23{\cdot}27 \times 232{\cdot}7}{1008{\cdot}4 \times 207}.$ **28.** $\dfrac{0{\cdot}078 \times 0{\cdot}0265}{0{\cdot}921 \times 1{\cdot}284}.$ **29.** $\dfrac{997 \times 0{\cdot}43}{22{\cdot}71 \times 1{\cdot}085}.$

30. $\sqrt{\dfrac{9{\cdot}347 \times 10{\cdot}73}{55{\cdot}09 \times 0{\cdot}173}}.$ **31.** $\sqrt[3]{\dfrac{1000}{3{\cdot}142 \times 372{\cdot}9}}.$ **32.** $\sqrt{\dfrac{27{\cdot}03 \times 0{\cdot}28}{(24{\cdot}48)^2}}.$

Exercise 26 j

Take $\pi = 3{\cdot}142$.

1. Use logarithms to evaluate to three significant figures $a\sqrt{b/c^3}$, where $a = 47{\cdot}2$, $b = 0{\cdot}3413$, $c = 0{\cdot}265$.

2. Use logarithms to find the value of $\sqrt{\dfrac{a(x^2 - y^2)}{bc}}$ when $a = 0{\cdot}3629$, $b = 15{\cdot}06$, $c = 0{\cdot}0259$, $x = 328{\cdot}374$, $y = 326{\cdot}259$.

3. Find c when $c^2 = 3082 - 2abx$, $a = 36{\cdot}2$, $b = 42{\cdot}1$, $x = 0{\cdot}7135$.

4. The time of oscillation (t sec.) of a pendulum l feet long is given by $t = 2\pi \sqrt{\dfrac{l}{32{\cdot}2}}$. Find the time of oscillation of a pendulum $8{\cdot}48$ feet long.

5. Through a gas main of diameter d inches and length L yards $1350 d^2 \sqrt{\dfrac{d}{0{\cdot}5L}}$ cubic feet of gas pass per hour. How many cubic feet per hour pass through a main $2\tfrac{1}{2}$ inches in diameter and a mile long?

6. Find A when $A = 2\pi r^2 + 2\pi rh$, $r = 0{\cdot}36$, $h = 19{\cdot}75$.

7. If $\dfrac{v^2}{r} = \dfrac{g}{289}$, calculate v when $r = 4000$, $g = \dfrac{32{\cdot}2}{5280}$. Also show that the value of $\dfrac{2\pi r}{v \times 60 \times 60}$ is approximately 24.

8. If $n = \dfrac{1}{2l}\sqrt{\dfrac{T}{4r^2}}$, find n when $l = 12{\cdot}3$, $T = 850000$, $r = 0{\cdot}14$.

9. The horse-power of the engines required to drive a ship of T tons at a speed of K knots may be calculated from the formula $0.0042 K^3 \times \sqrt[3]{T^2}$. Find, to three significant figures, the horse-power required for a ship of 30,700 tons which is to have a maximum speed of 25·4 knots.

10. The velocity (v ft. per sec.) of sound in air at temperature $t°$ C., when p lb. per sq. in. is the air pressure and d lb. per cu. ft. is the density of air under standard conditions, is given by

$$v = \sqrt{\left\{45 \cdot 08 \times \frac{p \times 144}{d} \times \left(1 + \frac{t}{273}\right)\right\}}.$$

Find v when $p = 14\cdot7$, $d = 0\cdot0809$, $t = 17$.

11. Calculate the value of x from the formula $x = \dfrac{4\pi^2 k l}{T^2 - t^2}$, having given $k = 0\cdot08974$, $l = 0\cdot202$, $T = 10\cdot18$, $t = 5\cdot804$.

12. The area of a triangle is given by the formula

$$A = \sqrt{s(s-a)(s-b)(s-c)}, \quad \text{where} \quad s = \tfrac{1}{2}(a+b+c).$$

Find to three significant figures the areas of triangles for which a, b, c are

(i) 3, 4, 4·5 in.   (ii) 6, 8, 9 cm.   (iii) 76·9, 93·1, 53·3 ft.

(iv) 0·927, 1·135, 0·675 metres.   (v) 85·6, 97·2, 105·4 yds.

13. Find to three significant figures the values of (i) $\sqrt{\dfrac{s(s-a)}{bc}}$, (ii) $\sqrt{\dfrac{s(s-b)}{ca}}$, where $s = \tfrac{1}{2}(a+b+c)$, $a = 25\cdot7$, $b = 33\cdot5$, $c = 30\cdot4$.

14. From a chemical experiment it was found that

$$V = 124 \times \frac{744}{760} \times \frac{273}{273+17} \quad \text{and} \quad \frac{D}{11160} - 0\cdot00129 = \frac{0\cdot3415}{V}.$$

Find D to three significant figures.

15. If $W = C^2 R + \dfrac{t^2}{R}$, find the value of W when $R = 0\cdot044$, $C = 41\cdot1$, $t = 1\cdot8$.

INDICES. LOGARITHMS I

16. A quantity Y is determined by the relation $Y = \dfrac{\text{stress}}{\text{strain}}$, where stress $= \dfrac{F}{\pi r^2}$ and strain $= \dfrac{x}{l}$. Find the value of Y when $F = 6000 \times 981$, $r = 0.0705$, $x = 0.147$, $l = 597$.

17. Evaluate $\dfrac{(1 \cdot 204)^4 - 1}{(1 \cdot 204)^4 + 1}$.

18.* Evaluate $\dfrac{(1 \cdot 03)^{10} - 1}{1 \cdot 03 - 1}$.

19.* Evaluate $\dfrac{748 \,[(1 \cdot 05)^{10} - 1]}{1 \cdot 05 - 1}$.

20. Evaluate $\sqrt{b^2 - 4ac}$ when $a = 1 \cdot 302$, $b = 2 \cdot 4$, $c = 1 \cdot 07$.

21. Evaluate $\dfrac{ab}{\sqrt{a^2 + b^2}}$ when $a = 15 \cdot 55$, $b = 18 \cdot 72$.

22. From the formula $t = 2\pi \sqrt{\dfrac{l}{g}}$, find l, correct to the nearest unit, when $t = 1 \cdot 20$ and $g = 981$.

23. Use the formula $V = \tfrac{4}{3}\pi r^3$ to find the radius of a sphere whose volume is 7296 cu. cm.

24. Use the formula $V = \tfrac{1}{12}\pi d^2 h$ to find the diameter of the base of a cone whose volume is 35 cu. ft. and height $3 \cdot 57$ ft.

25. If $s = \dfrac{1 \cdot 27 M}{d^2 h}$, find d to three decimal places when $s = 8 \cdot 9$, $M = 2 \cdot 51$, $h = 15$.

26. A rectangular block of lead whose dimensions, in inches, are $79 \cdot 12 \times 84 \cdot 64 \times 47 \cdot 21$, is melted and recast into 1000 equal cubes. Find, correct to three significant figures, the length of the edge of one of these cubes.

27. If $V = \dfrac{(t + 461) \times v}{T + 461}$, find t when $V = 770$, $v = 1050$, $T = 1200$.

* Use 7-figure logarithms for finding the powers. See end of book.

28. If $A = P\left(1 + \dfrac{r}{100}\right)^n$, (i) find P when $A = 5000$, $r = 5$, $n = 3$; (ii) find r when $A = 162 \cdot 1$, $P = 160$, $n = 2$.

29. The volume (V) of a cylindrical boiler with hemispherical ends is given by the formula $V = \pi r^2 l + \tfrac{4}{3}\pi r^3$, where l is the length of the cylindrical part and r is the radius.

Find the length of the cylindrical part if the volume of the boiler is 2775 cu. ft. and its radius is 5 ft. 3 in.

30. If $v = \pi l\,(R_1^2 - R_2^2)$, find R_1 when $R_2 = 13$, $l = 3 \cdot 6$, $v = 610$.

31. If $\dfrac{1}{x} = \dfrac{1}{u} + \dfrac{1}{v}$, find x when $u = 25 \cdot 24$, $v = 13 \cdot 27$.

32. If $\tfrac{4}{3}\pi p r^3 = 24 \cdot 2$ and $p = 7 \cdot 7$, calculate $4\pi r^2$.

33. If G gallons of water are delivered per hour by a pipe of diameter D inches and length L yards under a head of H feet, $G = \sqrt{\dfrac{(15D)^5 \times H}{L}}$; find the diameter of a pipe, two miles long, which will deliver 140,000 gallons of water per hour under a head of 30 feet.

34. What is the greatest integral power of 2 that is less than 10^8?

LOGARITHMS

143. A system of Indices.

In the previous chapter it was seen that by means of the graph of 2^x it is possible, within the limits of the graph, to express any number as a power of 2. This was confirmed algebraically.

For every number marked on the y axis and indicated on the graph, there is a corresponding index which can be read on the x axis. These constitute a system of indices by which numbers can be expressed as powers of a common basic number 2.

Similarly, by drawing graphs such as 3^x, 5^x, 10^x, numbers can be expressed as powers of 3, 5, or 10, or any other basic number.

Thus in all such cases it is possible to formulate systems of indices which, for any number A, would enable us to determine what power that number is of any other number B, which is called the base of the system.

This possibility of expressing any number as a power of any other number, and thus of the formation of a system of indices, as stated above, leads to practical results of great importance. It enables us to carry out, easily and accurately, calculations which would otherwise be laborious or even impossible. The fundamental ideas underlying this can be illustrated by means of a graph of powers similar to that drawn in Fig. 40. For this purpose we will use 10 as the base of the system and draw the graph of $y = 10^x$.

As powers of 10 increase rapidly, it will be possible to employ only small values for x, if the curve is to be of any use for our purpose. To obtain those powers we must use the rules for indices which were formulated in the previous chapter.

From Arithmetic we know that $\sqrt{10} = 3 \cdot 16$ app., *i.e.*, $10^{\frac{1}{2}} = 3 \cdot 16$.

Then $10^{0.25} = 10^{\frac{1}{4}} = (10^{\frac{1}{2}})^{\frac{1}{2}} = (3.16)^{\frac{1}{2}} = 1.78$ app. (by Arithmetic).
$10^{0.75} = 10^{\frac{3}{4}} = 10^{\frac{1}{2} + \frac{1}{4}} = 10^{\frac{1}{2}} \times 10^{\frac{1}{4}} = 3.16 \times 1.78$
$= 5.62$ app.
$10^{0.125} = 10^{\frac{1}{8}} = (10^{\frac{1}{4}})^{\frac{1}{2}} = (1.78)^{\frac{1}{2}} = 1.33$ app.

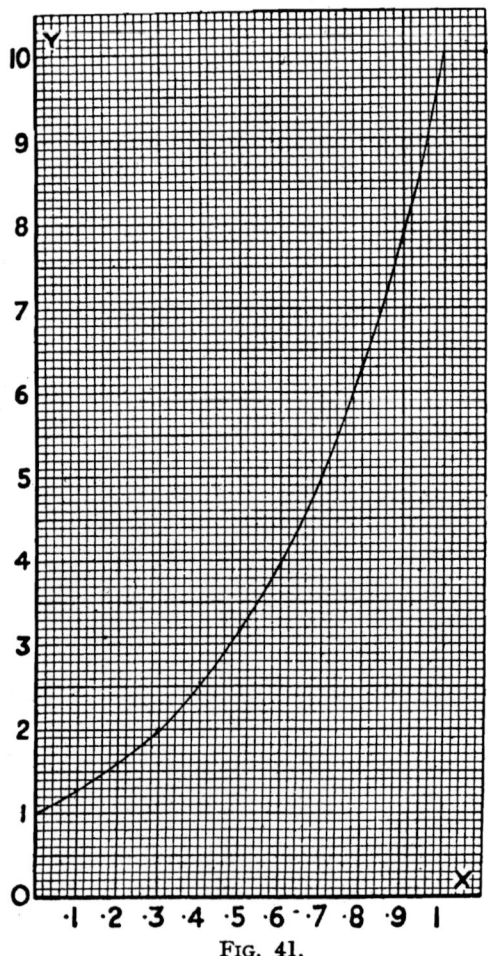

Fig. 41.

In this way a table of values for the curve can be compiled as follows:

LOGARITHMS

x	0	0·125	0·25	0·5	0·75	0·875	1
10^x	1	1·33	1·78	3·16	5·62	7·5	10

The resulting curve is shown in Fig. 41. The following examples illustrate the use that can be made of it in calculations.

Example 1. *Find from the graph the value of* $1·8 \times 2·6$.

From the graph
$$1·8 = 10^{0·26}$$
$$2·6 = 10^{0·42}$$
$$\therefore \quad 1·8 \times 2·6 = 10^{0·26} \times 10^{0·42}$$
$$= 10^{0·26 + 0·42} \quad \text{(first law of indices)}$$
$$= 10^{0·68}$$
$$= 4·6 \quad \text{from the graph.}$$

Example 2. *Find* $\sqrt[3]{9}$.

From the graph
$$9 = 10^{0·96}$$
$$\sqrt[3]{9} = 9^{\frac{1}{3}}$$
$$= (10^{0·96})^{\frac{1}{3}}$$
$$= 10^{0·32} \quad \text{(third law of indices)}$$
$$= 2·1 \quad \text{from the graph.}$$

144. A system of logarithms.

Although interesting as illustrating the principles involved, the above has little practical value for purposes of calculation, since we depend upon the readings from a curve which is necessarily limited in size and is not sufficiently accurate.

For practical purposes it is obvious that tables of the indices used, calculated to a suitable degree of accuracy, are necessary. Such tables have been compiled and are available for the purpose. For their compilation a more advanced knowledge of mathematics is required than is included in this volume.

The tables are constructed with 10 as a suitable base. They give the indices which indicate for all numbers, within the scope of the table, the powers they are of 10.

Such a table is called a **system of logarithms**, and the number 10, with respect to which the logarithms are calculated, is called the **base of the system**.

A logarithm to base 10 may be defined as follows:

The logarithm of a number to base 10 is the index of the power to which 10 must be raised to produce the number.

145. Notation for logarithms.

The student may wonder why another, and an unfamiliar term is employed as a name for an index. One reason for this will be seen from the following:

Let n be any positive number.
,, x be its index to base 10.
Then $n = 10^x$.

This is in reality a formula. If it is required to "*change the subject of the formula*" (see § 51) and express x in terms of the other letters, there is a difficulty in doing this concisely. Using words we could write:

$x =$ index of power of n to base 10.

This is cumbersome, so we employ the word "logarithm",[1] abbreviated to "log" as follows:

$$x = \log_{10} n$$

the number indicating the base being inserted as shown.

If the base is e, we write $x = \log_e n$.

In this form **x is expressed as a function of n**, whereas in the form $n = 10^x$, n is expressed as a function of x.

The student must be able to change readily from one form to another.

Examples.

(1) We saw in § 143 that $56 \cdot 2 = 10^{1 \cdot 75}$.

In log form this is $1 \cdot 75 = \log_{10} 56 \cdot 2$.

(2) $\qquad\qquad 1024 = 2^{10}$.
∴ $\qquad\qquad \log_2 1024 = 10$.

[1] The choice of the word logarithm can be explained only by the history of the word. The student could consult *A Short History of Mathematics*, by W. W. R. Ball.

LOGARITHMS

(3) $\qquad 1000 = 10^3.$
$\therefore \qquad \log_{10} 1000 = 3.$
(4) $\qquad 81 = 3^4.$
$\therefore \qquad \log_3 81 = 4.$

For ordinary calculations 10 is the most suitable base for a system of logarithms, but in more advanced mathematics a different base is required (see § 153).

146. Characteristic of a logarithm.

The integral or whole number part of a logarithm is called the characteristic. This can always be determined by inspection when logarithms are calculated to base 10, as will be seen from the following considerations:

Since
$10^0 = 1, \quad \log_{10} 1 = 0$
$10^1 = 10, \quad \log_{10} 10 = 1$
$10^2 = 100, \quad \log_{10} 100 = 2$
$10^3 = 1000, \quad \log_{10} 1000 = 3$
$10^4 = 10,000, \quad \log_{10} 10,000 = 4$

and so on.

From these results we see that,

for numbers between 1 and 10 the characteristic is 0
,, ,, ,, 10 ,, 100 ,, ,, ,, ,, 1
,, ,, ,, 100 ,, 1,000 ,, ,, ,, ,, 2
,, ,, ,, 1000 ,, 10,000 ,, ,, ,, ,, 3

and so on.

It is evident that the characteristic is always one less than the number of digits in the whole number part of the number.

Thus in $\log_{10} 3758 \cdot 7$ the characteristic is 3
$\log_{10} 375 \cdot 87$,, ,, ,, 2
$\log_{10} 37 \cdot 587$,, ,, ,, 1.

Thus the characteristics may always be determined by inspection, and consequently are not given in the tables. This is one advantage of having 10 for a base.

147. Mantissa of a logarithm.

The decimal part of a logarithm is called the mantissa.

In general the mantissa can be calculated to any required number of figures, by the use of higher mathematics. In most tables, such as those given in this volume, the mantissa is calculated to *four* places of decimals approximately. In *Chambers' " Book of Tables "* they are calculated to seven places of decimals.

The mantissa alone is given in the tables, and the following example will show the reason why:

$$\log_{10} 168{\cdot}3 = 2{\cdot}2261.$$
$$\therefore \quad 168{\cdot}3 = 10^{2{\cdot}2261}.$$
$$\therefore \quad 168{\cdot}3 \div 10 = 10^{2{\cdot}2261} \div 10^1.$$
$$\therefore \quad 16{\cdot}83 = 10^{2{\cdot}2261-1} \text{ (second law of indices)}$$
$$= 10^{1{\cdot}2261}.$$
$$\therefore \quad \log_{10} 16{\cdot}83 = 1{\cdot}2261.$$
Similarly $\log_{10} 1{\cdot}683 = 0{\cdot}2261$
and $\log_{10} 1683 = 3{\cdot}2261.$

Thus, if a number is multiplied or divided by a power of 10, the characteristic of the logarithm of the result is changed, but **the mantissa remains unaltered**. This may be expressed as follows:

Numbers having the same set of significant figures have the same mantissa in their logarithms.

148. To read a table of logarithms.

With the use of the above rules relating to the characteristic and mantissa of logarithms, the student should have no difficulty in reading a table of logarithms.

Below is a portion of such a table, giving the logarithms of numbers between 31 and 35.

No.	Log.	1	2	3	4	5	6	7	8	9	1	2	3	4	5	6	7	8	9
31	4914	4928	4942	4955	4969	4983	4997	5011	5024	5038	1	3	4	6	7	8	10	11	12
32	5051	5065	5079	5092	5105	5119	5132	5145	5159	5172	1	3	4	5	7	8	9	11	12
33	5185	5198	5211	5224	5237	5250	5263	5276	5289	5302	1	3	4	5	6	8	9	10	12
34	5315	5328	5340	5353	5366	5378	5391	5403	5416	5428	1	3	4	5	6	8	9	10	11
35	5441	5453	5465	5478	5490	5502	5514	5527	5539	5551	1	2	4	5	6	7	9	10	11

(1) (2)

The figures in column 1 in the complete table are the numbers from 1 to 99. The corresponding number in column 2 is the mantissa of the logarithm. As previously stated, the characteristic is not given, but can be written

LOGARITHMS

down by inspection. Thus $\log_{10} 31 = 1\cdot4914$, $\log_{10} 310 = 2\cdot4914$, etc.

If the number has a *third significant figure*, the mantissa will be found in the appropriate column of the next nine columns.

Thus $\log_{10} 31\cdot1 = 1\cdot4928$,
$\log_{10} 31\cdot2 = 1\cdot4942$, and so on.

If the number has a *fourth significant figure* space does not allow us to print the whole of the mantissa. But the next nine columns of what are called " mean differences " give us for every fourth significant figure a number which must be added to the mantissa already found for the first three significant figures. Thus if we want $\log_{10} 31\cdot67$, the mantissa for the first three significant figures 316 is 0·4997. For the fourth significant figure 7 we find in the appropriate column of mean differences the number 10. This is added to 0·4997 and so we obtain for the mantissa 5007.

$$\therefore \quad \log_{10} 31\cdot67 = 1\cdot5007.$$

Anti-logarithms.

The student is usually provided with a table of anti-logarithms which contains the *numbers corresponding to given logarithms*. These could be found from a table of logarithms but it is quicker and easier to use the anti-logarithms, which are given at the end of this book.

The tables are similar in their use to those for logarithms, but we must remember:

(1) That the mantissa of the log only is used in the table.

(2) When the significant figures of the number have been obtained, the student must proceed to fix the decimal point in them by using the rules which we have considered for the characteristic.

Example. *Find the number whose logarithm is* 2·3714.

First using the mantissa—viz., 0·3714—we find from the anti-logarithm table that the number corresponding is given as 2352. These are the first four significant figures of the number required.

Since the characteristic is 2, the number must lie between

100 and 1000 (see § 146) and therefore it must have 3 significant figures in the integral part.

∴ the number is 235·2.

Note.—As the log tables which will be usually employed by the beginner are all calculated to base 10, the base in further work will be omitted when writing down logarithms. Thus we shall write log 235·2 = 2·3714, the base 10 being understood.

Exercise 43.

1. Write down the characteristics of the logarithms of the following numbers:

    15,   1500,   31,672,   597,   8,   800,000
    51·63,.   3874·5,   2·615,   325·4

2. Read from the tables the logarithms of the following numbers:
    (1) 5, 50, 500, 50,000.
    (2) 4·7, 470, 47,000.
    (3) 52·8, 5·28, 528.
    (4) 947·8, 9·478, 94,780.
    (5) 5·738, 96·42, 6972.

3. Find, from the tables, the numbers of which the following are the logarithms:
    (1) 2·65, 4·65, 1·65.
    (2) 1·943, 3·943, 0·943.
    (3) 0·6734, 2·6734, 5·6734.
    (4) 3·4196, 0·7184, 2·0568.

149. Rules for the use of logarithms.

In using logarithms for calculations we must be guided by the laws which govern operations with them. Since logarithms are indices, these laws must be the same in principle as those of indices. These laws are given below; formal proofs are omitted. They follow directly from the corresponding index laws.

(1) Logarithm of a product.

The logarithm of the product of two or more numbers is equal to the sum of the logarithms of these numbers (see first law of indices).

LOGARITHMS

Thus if p and q be any numbers

$$\log (p \times q) = \log p + \log q.$$

(2) **Logarithm of a quotient.**

The logarithm of p divided by q is equal to the logarithm of p diminished by the logarithm of q (see second law of indices).

Thus $\log (p \div q) = \log p - \log q.$

(3) **Logarithm of a power.**

The logarithm of a power of a number is equal to the logarithm of the number multiplied by the index of the power (see third law of indices).

Thus $\log a^n = n \log a.$

(4) **Logarithm of a root.**

This is a special case of the above (3).

Thus
$$\log \sqrt[n]{a} = \log a^{\frac{1}{n}}$$
$$= \frac{1}{n} \log a.$$

150. Examples of the use of logarithms.

Example I. *Find the value of* $57{\cdot}86 \times 4{\cdot}385$.

Let $x = 57{\cdot}86 \times 4{\cdot}385.$
Then $\log x = \log 57{\cdot}86 + \log 4{\cdot}385$
$= 1{\cdot}7624 + 0{\cdot}6420$
$= 2{\cdot}4044$
$= \log 253{\cdot}7.$
$\therefore \quad x = 253{\cdot}7.$

No.	log.
57·86	1·7624
4·385	0·6420
253·7	2·4044

Notes.—(1) The student should remember that the logs in the tables are correct to four significant figures only. Consequently he cannot be sure of four significant figures in the answer. It would be more correct to give the above answer as 254, correct to three significant figures.

(2) The student is advised to adopt some systematic way of arranging the actual operations with logarithms. Such a method is shown above.

TEACH YOURSELF ALGEBRA

Example 2. *Find the value of*
$$\frac{5 \cdot 672 \times 18 \cdot 94}{1 \cdot 758}.$$

Let $x = \dfrac{5 \cdot 672 \times 18 \cdot 94}{1 \cdot 758}$.

$\therefore \log x = \log 5 \cdot 672 + \log 18 \cdot 94 - \log 1 \cdot 758$
$\quad\quad\quad = 0 \cdot 7538 + 1 \cdot 2774 - 0 \cdot 2450$
$\quad\quad\quad = 1 \cdot 7862$
$\quad\quad\quad = \log 61 \cdot 12.$

$\therefore \quad x = 61 \cdot 12$
or $\quad\quad x = 61 \cdot 1$ (to three significant figures).

No.	log.
5·672	0·7538
18·94	1·2774
	2·0312
1·758	0·2450
61·12	1·7862

Example 3. *Find the fifth root of* 721·8.

Let $\quad\quad\quad x = \sqrt[5]{721 \cdot 8}$
$\quad\quad\quad\quad\quad = (721 \cdot 8)^{\frac{1}{5}}.$
Then $\quad\quad \log x = \frac{1}{5} \log 721 \cdot 8 \quad$ (see § 149(4))
$\quad\quad\quad\quad\quad = \frac{1}{5}(2 \cdot 8584)$
$\quad\quad\quad\quad\quad = 0 \cdot 5717.$
$\therefore \quad\quad\quad x = 3 \cdot 730.$

Exercise 44.

Use logarithms to find the values of the following:

1. $23 \cdot 4 \times 14 \cdot 73$.
2. $43 \cdot 97 \times 6 \cdot 284$.
3. $987 \cdot 4 \times 1 \cdot 415$.
4. $42 \cdot 7 \times 9 \cdot 746 \times 14 \cdot 36$.
5. $28 \cdot 63 \div 11 \cdot 95$.
6. $43 \cdot 97 \div 6 \cdot 284$.
7. $23 \cdot 4 \div 14 \cdot 73$.
8. $927 \cdot 8 \div 4 \cdot 165$.
9. $94 \cdot 76 \times 4 \cdot 195 \div 27 \cdot 94$.
10. $\dfrac{15 \cdot 36 \times 9 \cdot 47 \times 11 \cdot 48}{5 \cdot 632 \times 21 \cdot 85}$.
11. $(9 \cdot 478)^3$.
12. $(51 \cdot 47)^2$.
13. $(1 \cdot 257)^5$.
14. $(15 \cdot 23)^2 \times 3 \cdot 142$.
15. $(5 \cdot 98)^2 \div 16 \cdot 47$.
16. $\dfrac{(91 \cdot 5)^2}{4 \cdot 73 \times 16 \cdot 92}$.
17. $\dfrac{(8 \cdot 97)^2 \times (1 \cdot 059)^3}{57 \cdot 7}$.
18. $\dfrac{4798}{(56 \cdot 2)^2 \div (9 \cdot 814)^3}$.
19. $\sqrt[4]{3 \cdot 417}$.
20. $\sqrt[3]{4 \cdot 872}$.
21. $\sqrt[3]{1 \cdot 625^2 \times 4 \cdot 738}$.
22. $\sqrt[5]{61 \cdot 5 \times 2 \cdot 73}$.

23. If $\pi r^2 = 78 \cdot 6$ find r when $\pi = 3 \cdot 142$.
24. If $\frac{4}{3}\pi r^3 = 15 \cdot 5$, find r when $\pi = 3 \cdot 142$.

LOGARITHMS

151. Logarithms of numbers between 0 and 1.

In § 146 we gave examples of powers of 10 when the index is a positive integer. We will now consider cases in which the indices are negative.

Thus $10^1 = 10$ $\therefore \log_{10} 10 = 1$
$10^0 = 1$ $\therefore \log_{10} 1 = 0$
$10^{-1} = \frac{1}{10} = 0.1$ $\therefore \log_{10} 0.1 = -1$
$10^{-2} = \frac{1}{10^2} = 0.01$ $\therefore \log_{10} 0.01 = -2$
$10^{-3} = \frac{1}{10^3} = 0.001$ $\therefore \log_{10} 0.001 = -3$
             etc.

From these results we may deduce that:
The logarithms of numbers between 0 and 1 are always negative.

We have seen (§ 147) that if a number be divided by 10, we obtain the log of the result by subtracting 1.

Thus if   log 49·8   = 1·6972
      log 4·98   = 0·6972
      log 0·498   = 0·6972 — 1
      log 0·0498 = 0·6972 — 2
      log 0·00498 = 0·6972 — 3.
From the above, log 0·498 = 0·6972 — 1
          = — 0·3028.

Now, in the logs of numbers greater than unity, the mantissa remains the same when the numbers are multiplied or divided by powers of 10 (see § 147), *i.e.* with the same significant figures we have the same mantissa.

It would clearly be a great advantage if we could find a system which would enable us to use this rule for numbers less than unity, and so avoid, for example, having to write

    log 0·498 as — 0·3028.

This can be done by not carrying out the subtraction as shown above, and writing down the characteristic as negative. But to write log 0·498 as 0·6972 — 1 would be awkward. Accordingly we adopt the notation $\bar{1}\cdot6972$, writing the minus sign above the characteristic.

It is very important to remember that
$$\bar{1}\cdot 6972 = -1 + 0\cdot 6972.$$

Thus in logarithms written in this way **the charactistic is negative and the mantissa is positive.**

With this notation log 0·0498 = $\bar{2}$·6972
log 0·00498 = $\bar{3}$·6972
log 0·000498 = $\bar{4}$·6972   etc.

Note.—The student should note that *the negative characteristic is numerically one more than the number of zeros after the decimal point.*

Example 1. *From the tables find the logs of* 0·3185, 0·03185 *and* 0·003185.

Using the portion of the tables in § 148, we see that the mantissa for 0·3185 will be 0·5031.

Also the characteristic is — 1.

∴ log 0·3185 = $\bar{1}$·5031.
Similarly log 0·03185 = $\bar{2}$·5031
and log 0·003185 = $\bar{3}$·5031.

Example 2. *Find the number whose log is* $\bar{3}$·5416.

From the anti-log tables we find that the significant figures of the number whose mantissa is 5416 are 3480. As the characteristic is — 3, there will be two zeros after the decimal point.

∴ the number is   **0·003480**

(correct to 4 significant figures).

Exercise 45.

1. Write down the logarithms of:
    (1) 2·798, 0·2798, 0·02798.
    (2) 4·264, 0·4264, 0·004264.
    (3) 0·009783, 0·0009783, 0·9783.
    (4) 0·06451, 0·6451, 0·0006451.

2. Write down the logarithms of:
    (1) 0·05986.   (4) 0·00009275.
    (2) 0·000473.   (5) 0·5673.
    (3) 0·007963.   (6) 0·07986.

LOGARITHMS

3. Find the numbers whose logarithms are:
   - (1) $\bar{1}\cdot3342$.
   - (2) $\bar{3}\cdot8724$.
   - (3) $\bar{2}\cdot4871$.
   - (4) $\bar{4}\cdot6437$.
   - (5) $\bar{1}\cdot7738$.
   - (6) $\bar{8}\cdot3948$.

152. Operations with logarithms which are negative.

Care is needed in operating with the logarithms of numbers which lie between 0 and 1, since they are negative and, as shown above, are written with the characteristic negative and the mantissa positive.

A few examples will show the method of working.

Example 1. *Find the sum of the logarithms :*

$\bar{1}\cdot6173$, $\bar{2}\cdot3415$, $\bar{1}\cdot6493$, $0\cdot7374$.

Arranging thus
$\bar{1}\cdot6173$
$\bar{2}\cdot3415$
$\bar{1}\cdot6493$
$0\cdot7374$

$\overline{\bar{2}\cdot3455}$

The point to be specially remembered is that the 2 which is carried forward from the addition of the mantissæ is positive, since they are positive. Consequently the addition of the characteristics becomes

$$-1-2-1+0+2=-2.$$

Example 2. *From the logarithm $\bar{1}\cdot6175$ subtract the log $\bar{3}\cdot8463$.*

$\bar{1}\cdot6175$
$\bar{3}\cdot8463$

$\overline{1\cdot7712}$

Here in " borrowing " to subtract the 8 from the 6, the -1 in the top line becomes -2, consequently on subtracting the characteristics we have

$$-2-(-3)=-2+3=+1.$$

Example 3. *Multiply $\bar{2}\cdot 8763$ by 3.*

$$\begin{array}{r}\bar{2}\cdot 8763 \\ 3 \\ \hline \bar{4}\cdot 6289\end{array}$$

From the multiplication of the mantissa, 2 is carried forward. But this is positive and as $(-2) \times 3 = -6$, the characteristic becomes $-6 + 2 = -4$.

Example 4. *Multiply $\bar{1}\cdot 8738$ by $1\cdot 3$.*

In a case of this kind it is better to multiply the characteristic and mantissa separately and add the results.

Thus $\qquad 0\cdot 8738 \times 1\cdot 3 = 1\cdot 13594$
$\qquad\qquad\quad -1 \times 1\cdot 3 = -1\cdot 3$.

$-1\cdot 3$ is wholly negative and so we change it to $\bar{2}\cdot 7$, to make the mantissa positive.

Then the product is the sum of

$$\begin{array}{r}1\cdot 13594 \\ \bar{2}\cdot 7 \\ \hline \bar{1}\cdot 83594\end{array}$$

or $\qquad\qquad 1\cdot 8359$ approx.

Example 5. *Divide $\bar{5}\cdot 3716$ by 3.*

Here the difficulty is that on dividing $\bar{5}$ by 3 there is a remainder 2 which is negative, and cannot therefore be carried on to the positive mantissa. To get over the difficulty we write:

$\qquad\qquad\qquad -5 = -6 + 1$
or the log as $\qquad -6 + 1\cdot 3716$.

Then the division of the -6 gives us -2 and the division of the positive part $1\cdot 3716$ gives $0\cdot 4572$, which is positive. Thus the complete quotient is $\bar{2}\cdot 4572$. The work might be arranged thus:

$$\begin{array}{r}3)\bar{6} + 1\cdot 3716 \\ \hline \bar{2} + 0\cdot 4572 \\ \hline = \bar{2}\cdot 4572\end{array}$$

LOGARITHMS

Exercise 46.

1. Add together the following logarithms:
   - (1) $\bar{2}\cdot5178 + 1\cdot9438 + 0\cdot6138 + \bar{5}\cdot5283.$
   - (2) $3\cdot2165 + \bar{3}\cdot5189 + \bar{1}\cdot3297 + \bar{2}\cdot6475.$

2. Find the values of:
   - (1) $4\cdot2183 - 5\cdot6257.$
   - (2) $0\cdot3987 - \bar{1}\cdot5724.$
   - (3) $\bar{1}\cdot6472 - \bar{1}\cdot9875.$
   - (4) $\bar{2}\cdot1085 - \bar{5}\cdot6271.$

3. Find the values of:
   - (1) $\bar{1}\cdot8732 \times 2.$
   - (2) $\bar{2}\cdot9456 \times 3.$
   - (3) $\bar{1}\cdot5782 \times 5.$
   - (4) $\bar{1}\cdot5782 \times 1\cdot5.$
   - (5) $\bar{2}\cdot9947 \times 0\cdot8.$
   - (6) $\bar{2}\cdot7165 \times 2\cdot5.$

4. Find the values of:
   - (1) $\bar{3}\cdot9778 \times 0\cdot65.$
   - (2) $\bar{2}\cdot8947 \times 0\cdot84.$
   - (3) $\bar{1}\cdot6257 \times 0\cdot6.$
   - (4) $2\cdot1342 \times -0\cdot4.$
   - (5) $1\cdot3164 \times -1\cdot5.$
   - (6) $\bar{1}\cdot2976 \times -0\cdot8.$

5. Find the values of:
   - (1) $\bar{1}\cdot4798 \div 2.$
   - (2) $\bar{2}\cdot5637 \div 5.$
   - (3) $\bar{4}\cdot3178 \div 3.$
   - (4) $\bar{3}\cdot1195 \div 2.$
   - (5) $\bar{1}\cdot6173 \div 1\cdot4.$
   - (6) $\bar{2}\cdot3178 \div 0\cdot8.$

---

153. Change of base of a system of logarithms.

Although logs calculated to base 10 are usually employed for calculations, in more advanced Mathematics, as well as in Engineering, the logs which naturally arise are calculated to a base which is given by the series

$$1 + \frac{1}{1} + \frac{1}{1\cdot2} + \frac{1}{1\cdot2\cdot3} + \frac{1}{1\cdot2\cdot3\cdot4} + \ldots \text{ to infinity.}$$

This series is denoted by **e**, and its value can be calculated to any required degree of accuracy by taking sufficient terms. To 5 places of decimals, $e = 2\cdot71828$.

Logs calculated to this base are called **Naperian logarithms**, after Lord Napier, who discovered them in 1614, using this base. They are also called **Natural logarithms** or **Hyperbolic logarithms**.

TEACH YOURSELF ALGEBRA

The student, possessing only tables of logs to base 10 may require to use the logs of numbers to base e, and must therefore know how to find them.

The relations between the logs of numbers to different bases is found as follows:

Let n be any number.
Let a and b be two bases.

Suppose that logs to base b are known, and we require to find them to base a.

Let $\log_b n = y$, $\therefore n = b^y$.
Then $\log_a n = \log_a (b^y)$
$= y \log_a b.$ (§ 149, rule 3.)
$\therefore \log_a n = \log_b n \times \log_a b.$

Thus, knowing the log of a number to a base b, we find its log to base a by multiplying, whatever the number, by $\log_a b$.

In the above result let $b = 10$ and $a = e$.

Then $\log_e n = \log_{10} n \times \log_e 10.$

In this result let $n = e$
then $\log_e e = \log_{10} e \times \log_e 10$
but $\log_e e = 1.$
$\therefore \log_{10} e \times \log_e 10 = 1.$
$\therefore \log_e 10 = \dfrac{1}{\log_{10} e}.$

$\therefore$ in the rule $\log_e n = \log_{10} n \times \log_e 10$
we can write $\log_e n = \log_{10} n \times \dfrac{1}{\log_{10} e}.$

Thus both logs on the right-hand side are to base 10.

Now $\log_{10} e = 0\cdot 4343$
and $\log_e 10 = \dfrac{1}{0\cdot 4343} = 2\cdot 2036.$

Hence to change from base 10 to base e, we may use either of the following:

(1) $\log_e n = \log_{10} n \times 2\cdot 3026$
or (2) $\log_e n = \log_{10} n \div 0\cdot 4343.$

LOGARITHMS

Example. *Find* $\log_e 50$.

Using $\log_e 50 = \log_{10} 50 \times 2\cdot 3026$
we have $\log_e 50 = 1\cdot 6990 \times 2\cdot 3026$.

Evaluating the right-hand side by use of logs we get

$$\log_e 50 = 3\cdot 913.$$

Summary of the laws relating to logarithms, together with some special points the truth of which will be obvious.

(1) The logarithm of the base itself is always unity.
(2) The logarithm of 1 is always zero, whatever the base.
(3) The logarithms of all numbers less than unity are negative.
(4) The logarithm of a number is equal to — (the log of its reciprocal).

Thus $\log_a n = -\log_a \dfrac{1}{n}$.

(5) $\log_a (x \times y) = \log_a x + \log_a y$.
(6) $\log_a (x \div y) = \log_a x - \log_a y$.
(7) $\log_a x^n = n \log_a x$.
(8) $\log_a \sqrt[n]{x} = \dfrac{1}{n} \log_a x$.

Exercise 47.

Miscellaneous Exercises in the Use of Logarithms.

1. $15\cdot 62 \times 0\cdot 987$.
2. $0\cdot 4732 \times 0\cdot 694$.
3. $0\cdot 513 \times 0\cdot 0298$.
4. $75\cdot 94 \times 0\cdot 0916 \times 0\cdot 8194$.
5. $9\cdot 463 \div 15\cdot 47$.
6. $0\cdot 9635 \div 29\cdot 74$.
7. $27\cdot 91 \div 569\cdot 4$.
8. $0\cdot 0917 \div 0\cdot 5732$.
9. $5\cdot 672 \times 14\cdot 83 \div 0\cdot 9873$.
10. $(0\cdot 9173)^2$.
11. $(0\cdot 4967)^3$.
12. $\sqrt[3]{1\cdot 715}$.
13. $\sqrt[5]{647\cdot 2} \div (3\cdot 715)^3$.
14. $\frac{1}{2}(48\cdot 62)^4$.
15. $\sqrt[3]{\dfrac{9\cdot 728}{3\cdot 142}}$.
16. $(1\cdot 697)^{2\cdot 4}$.
17. $(19\cdot 72)^{0\cdot 57}$.
18. $(0\cdot 478)^{3\cdot 1}$.
19. $(5\cdot 684)^{-1\cdot 12}$.
20. $(0\cdot 5173)^{-3\cdot 4}$.
21. $\sqrt[4]{0\cdot 01697}$.
22. $(0\cdot 1478)^2 \div 0\cdot 6982$.
23. $\sqrt[3]{0\cdot 8172} \div \sqrt[4]{0\cdot 7658}$.

TEACH YOURSELF ALGEBRA

24. $9.74^2 - 5.66^2$. (Hint.—This should first be factorized and changed to a product.)

25. $\dfrac{9.32 \times 0.761}{\sqrt{18.2}}$. **26.** $\sqrt[3]{\dfrac{647.3 \times 3.2}{3.142 \times 10.78}}$.

27. $\sqrt{(3.62)^2 + (5.47)^2 + (6.91)^2}$.

28. Find the value of πr^2 when $\pi = 3.142$ and $r = 16.89$.

29. Find the value of $\tfrac{4}{3}\pi r^3$ when $\pi = 3.142$ and $r = 2.9$.

30. If $V = pv^{1.6}$ find V when $v = 6.032$ and $p = 29.12$.

31. If $3^x = 24$, find x.

32. If $R^n = 1.8575$ find R when $n = 18$.

33. When $l = \dfrac{gt^2}{\pi^2}$ find l when $g = 32.2$, $t = 2$, $\pi = 3.142$.

34. From the formula $V = \sqrt{\dfrac{2ghD}{0.03L}}$ find V when $g = 32.2$, $h = 0.627$, $L = 175$, $D = 0.27$.

35. If $t = 2\pi\sqrt{\dfrac{l}{g}}$, find g when $l = 5.304$, $t = 2.55$, $\pi = 3.142$.

36. Without using tables find the values of:

(a) $\log 27 \div \log 3$.   (b) $(\log 16 - \log 2) \div \log 2$.

37. If $M = PR^n$ find M when $P = 200$, $R = 1.05$, $n = 20$.

38. Find the radius of a sphere whose volume is 500 c. ft. $V = \tfrac{4}{3}\pi r^3$.

39. Find the values of (1) $\log_e 4.6$
(2) $\log_e 0.062$.

40. The insulating resistance, R, of a wire of length l is given by
$$R = \dfrac{0.42S}{l} \times \log_e \dfrac{d_2}{d_1}.$$

Find l when $S = 2000$, $R = 0.44$, $d_2 = 0.3$, $d_1 = 0.16$.

41. In a calculation on the dryness of steam the following formula was used:
$$\dfrac{qL}{T} = \dfrac{q_1 L_1}{T_1} + \log_e \dfrac{T_1}{T}.$$

Find q when $L_1 = 850$, $L = 1000$, $T_1 = 780$, $T = 650$, $q_1 = 1$.

Exercise 43.

1. 1, 3, 4, 2, 0, 5, 1, 3, 0, 2.
2. (1) 0·6990, 1·6990, 2·6990, 4·6990.
   (2) 0·6721, 2·6721, 4·6721.
   (3) 1·7226, 0·7226, 2·7226.
   (4) 2·9767, 0·9767, 4·9767.
   (5) 0·7588, 1·9842, 3·8433.
3. (1) 446·7, 44670, 44·67.   (2) 87·70, 8770, 8·770.
   (3) 4·714, 471·4, 471,400.   (4) 2628, 5·229, 114·0.

Exercise 44.

1. 344·6.	2. 276·4.	3. 1397.
4. 5975.	5. 2·396.	6. 6·997.
7. 1·589.	8. 222·8.	9. 14·22.
10. 13·56.	11. 851·3.	12. 2650.
13. 3·137.	14. 728·5.	15. 2·172.
16. 104·6.	17. 1·656.	18. 1436.
19. 1·359.	20. 1·695.	21. 2·321.
22. 2·786.	23. 5·002.	24. 1·516.

TEACH YOURSELF ALGEBRA

Exercise 45.
1. (1) 0·4470, $\bar{1}$·4470, $\bar{2}$·4470. (2) 0·6298, $\bar{1}$·6298, $\bar{3}$·6298.
   (3) $\bar{3}$·9904, $\bar{4}$·9904, $\bar{1}$·9904. (4) $\bar{2}$·8097, $\bar{1}$·8097, $\bar{4}$·8097.
2. (1) $\bar{2}$·7771. (2) $\bar{4}$·6749. (3) $\bar{3}$·9011.
   (4) $\bar{5}$·9673. (5) $\bar{1}$·7538. (6) $\bar{2}$·9023.
3. (1) 0·2159. (2) 0·007453. (3) 0·03070.
   (4) 0·0004402. (5) 0·5940. (6) $2·482 \times 10^{-6}$.

Exercise 46.
1. (1) $\bar{4}$·6037. (2) $\bar{2}$·7126.
2. (1) $\bar{2}$·5926. (2) 0·8263. (3) $\bar{1}$·6597.
   (4) 2·4814.
3. (1) $\bar{1}$·7464. (2) $\bar{4}$·8368. (3) $\bar{3}$·8910.
   (4) $\bar{1}$·3673. (5) $\bar{1}$·1958. (6) $\bar{4}$·7913.
4. (1) $\bar{2}$·6856. (2) $\bar{1}$·07155. (3) $\bar{1}$·7754.
   (4) $\bar{1}$·1463. (5) $\bar{2}$·0254. (6) 0·5619.
5. (1) $\bar{1}$·7399. (2) $\bar{1}$·7127. (3) $\bar{2}$·7726.
   (4) $\bar{2}$·5598. (5) $\bar{1}$·7266. (6) $\bar{3}$·8973.

LOGARITHMS

267. Definitions. If we consider the equation $a^x = y$, the problem, given two of the three numbers, a, x, y, to find the third, leads to the consideration of the following types:
1. Given x and y, to find a.
2. Given a and x, to find y.
3. Given a and y, to find x.

The solution of the first presents itself in the form $a = \sqrt[x]{y}$, by taking the xth root of both members of the equation.

The problem in the second is to raise a to the xth power, which operation is called *exponentiation*. In this operation a is called the *base* and a^x the exponential of x with regard to the base a.

In the third case the operation of finding x when a and y are given is called the *logarithmetic* operation, and is expressed in symbols by the equation $x = \log_a y$. From this it is seen that the equation $a^x = y$ may be written in the form $a^{\log_a y} = y$. **The logarithm** *of a number y to the base " a " is that exponent which indicates the power to which the base must be raised in order to produce y.* The expression $\log_a y$ is read "logarithm of y with respect to the base a." It is seen that $\log_a y$, when found, is simply an exponent, and as such is subject to the laws of indices.

268. Since $a^0 = 1$ for all finite values of a different from zero, it follows that $\log_a 1 = 0$. Since $a^1 = a$, it follows that $\log_a a = 1$.

Since $a^{+\infty} = \infty$, and $a^{-\infty} = 0$, for all finite values of $a > 1$, it follows for such values of a, that $\log_a 0 = -\infty$ and $\log_a \infty = +\infty$.

Similarly for all positive values of $a < 1$, $\log_a 0 = \infty$ and $\log_a \infty = -\infty$.

If $a > 0$ and x is a real number, a^x cannot be negative, therefore the logarithm of a negative number is not real.

If $a^x = m$, $a^y = n$, and $m > n$, that is, if $a^x > a^y$, it is obvious that $x > y$ when $a > 1$, that is, when the base is greater than unity, the greater the number the greater the logarithm, and conversely.

269. Theorems. Let $a^x = m$, $a^y = n$, then $mn = a^x a^y = a^{x+y}$.

Therefore, $\log_a(mn) = x + y = \log_a m + \log_a n$.

Thus, if $\log_{10} 2 = 0.3010$ and $\log_{10} 3 = 0.4771$,

then $\log_{10} 6 = \log_{10} 2 + \log_{10} 3 = 0.7781$.

Similarly,
$$\log_a(mn \cdots p) = \log_a(mn \cdots) + \log_a p$$
$$= \log_a m + \log_a n + \cdots + \log_a p,$$

that is, *the logarithm of a product is equal to the sum of the logarithms of its factors.*

Again, $\dfrac{m}{n} = \dfrac{a^x}{a^y} = a^{x-y}$.

Therefore, $\log_a\left(\dfrac{m}{n}\right) = x - y = \log_a m - \log_a n$,

that is, *the logarithm of a quotient is equal to the logarithm of the dividend minus the logarithm of the divisor.*

Thus
$\log_{10} 5 = \log_{10} 10 - \log_{10} 2 = 1.0000 - 0.3010 = 0.6970$.

LOGARITHMS

Raising both sides of the equation $a^x = m$ to the kth power, where k is any real number, integral or fractional, positive or negative, we have $a^{kx} = m^k$.

Therefore, by definition, $\log_a (m^k) = k \cdot x = k \log_a m$, that is, *the logarithm of any power of a number is the logarithm of the number multiplied by the index of the power, whether the index be integral or fractional, positive or negative.*

Thus $\quad \log_{10} 8 = \log_{10} 2^3 = 3 \log_{10} 2 = 0.9030,$

and $\quad \log_{10} \sqrt[3]{2} = \log_{10} 2^{\frac{1}{3}} = \tfrac{1}{3} \log_{10} 2 = 0.1003.$

NOTE. Since the remainder, when divided by the divisor, gives a quotient which is less than one half, it is neglected; if that quotient were greater than one half, it would be called *unity*.

For a base $a > 1$, $\log_a \dfrac{p}{q} = \log_a p - \log_a q$, which is positive or negative according as $p \gtreqless q$; that is, the logarithm of a number greater than unity is positive and of a number less than unity is negative.

If $1 > a > 0$ and $a^x > a^y$, then $\left(\dfrac{1}{a}\right)^y > \left(\dfrac{1}{a}\right)^x$ and therefore $y > x$, and the greater the number the less the logarithm, and conversely. The $\log_a \dfrac{p}{q} = \log_a p - \log_a q$, which is positive or negative according as $p \lesseqgtr q$. Therefore with this base the logarithm of a number greater than unity is negative and of a number less then unity is positive.

We have seen, therefore, that *for a base greater than unity the logarithm of a number greater than unity is positive and of a number less than unity is negative; while for a base less than unity the logarithm of a number greater than unity is negative and of a number less than unity is positive.* The same result may be arrived at by substituting $\dfrac{1}{b}$ for b in the transformation formula of **271**.

COLLEGE ALGEBRA

270. **EXAMPLES**

1. If a series of numbers are in G. P. their logarithms are in A. P.
2. Given $2^x = 8$, find x.
3. Given $\log_3 27 = x$, find x.
4. Given $\log_x 32 = 5$, find x.

Given $\log 2 = 0.3010$, $\log 3 = 0.4771$, $\log 7 = 0.8451$; find the logarithms of the following numbers to the same base:

5. 14.   8. 32.   11. $14\frac{2}{7}$.   14. $\sqrt[3]{96}$.

6. 28.   9. $10\frac{1}{2}$.   12. 2.31.   15. $\sqrt[7]{48}$.

7. 24.   10. $20\frac{4}{7}$.   13. $\sqrt{56}$.   16. $\sqrt[5]{13.5}$.

17. Find the logarithm of 243 to the base 9.

Let x be the required logarithm.

Then $\qquad 9^x = 243$.

But 9 and 243 are both powers of 3. Hence $3^{2x} = 3^5$,
and $\qquad 2x = 5$, or $x = 2.5$.

18. Find the logarithm of 32 to the base 4.
19. Find the logarithm of 4 to the base 32.
20. Find the logarithm of $\frac{1}{343}$ to the base 49.
21. Find the logarithm of 343 to the base $\frac{1}{49}$.
22. Find the logarithm of $\frac{1}{27}$ to the base $\frac{1}{9}$.

271. Let $\log_a n = x$, $\log_b n = y$, then
$$a^x = n \text{ and } b^y = n$$
and therefore $\qquad a^x = b^y$.

LOGARITHMS

Taking the logarithm of both members of the last equation with respect to any third base c, we have

$$x \log_c a = y \log_c b,$$

or by substituting the values of x and y,

$$\log_a n \log_c a = \log_b n \log_c b,$$

whence $\log_b n = \dfrac{\log_c a}{\log_c b} \log_a n.$

Since c was any real number whatever, the ratio $\dfrac{\log_c a}{\log_c b}$ is constant for any given value of a and b, and is called the *modulus of the transformation*.

For $c = a$, we have $\dfrac{\log_c a}{\log_c b} = \dfrac{1}{\log_a b}$,

and $\log_b n = \dfrac{1}{\log_a b} \log_a n.$

This is a formula for transforming logarithms of numbers which are known with respect to a base a into logarithms of those numbers with respect to any other base b.

The student may show that $\log_a b \, \log_b a = 1$.

272. Although theoretically any positive number except unity could be made the base of a system of logarithms, yet for practical purposes only two systems are at all frequently used. One, called the *natural system*, or sometimes the *Napierian system*, is explained in a subsequent article, **283**. The other, known as the *Briggs*, or *common, system*, has the number 10 for its base. It is used for all purposes involving merely numerical calculation. The advantage of this base consists in the fact that any change in the position of the decimal point in a number will merely add an integer to, or subtract it from, the logarithm, because the number will then merely be multiplied or divided by an integral power of 10, **269**. Hence the fractional portion of the logarithm will be the same for any sequence of figures, whatever the position of the decimal point among them.

273. In the Briggs system the logarithm of 1 is 0, and the logarithm of 10 is 1, by **268**. Hence the logarithm of any number between 1 and 10 is a positive fraction. Every number greater than 10 may be obtained by multiplying a number between 1 and 10 by an integral power of 10. Hence its logarithm consists of an integer plus a fraction. This integer, which has been shown to be dependent merely upon the position of the decimal point in the number, is called the *characteristic of the logarithm*. The fractional part, which depends merely upon the sequence of figures in the number, and is ordinarily written in the form of a decimal, is called the *mantissa of the logarithm*. The mantissa is ordinarily not a terminating decimal, but is carried out four, five, six, etc. places according to the degree of accuracy required in the work.

Thus log $2 = 0.3010$, and log $200 = 2.3010$. In each case the mantissa is .3010, while the characteristics are, respectively, 0 and 2.

274. The logarithm of a positive number less than unity is really negative, but since such a number may be derived from a number between 1 and 10 by dividing it by an integral power of 10, it is convenient to regard the logarithm as composed of a positive mantissa and a negative characteristic. Thus if log $2 = 0.3010$, then since $.002 = \frac{2}{10^3}$, log $.002 = 0.3010 - 3$.

275. Two methods are in common use for writing logarithms with negative characteristics. Thus log $.002 = \bar{3}.3010$, where the negative sign is placed over the characteristic to indicate that it alone is negative; or the -3 is called $7 - 10$

LOGARITHMS

and the logarithm is written $7.3010 - 10$. By this means the negative portion is always a multiple of 10, and is kept quite separate from the positive portion of the logarithm.

276. To find the logarithm of a given number.

1. *To find the characteristic.* In **273** we have seen that if the decimal point follows the first significant digit, the characteristic of the logarithm is 0. For every place the decimal point in the number is moved toward the right the number is multiplied by 10. Hence if the number is greater than 10, the characteristic is one less than the number of significant figures to the left of the decimal point. Likewise for every place the decimal point is moved toward the left the number is divided by 10. Hence if the decimal point immediately precedes the first significant digit, the characteristic of the logarithm is -1, or $9-10$. If one cipher intervenes, it is $8-10$, and so on, subtracting one for each additional cipher.

The characteristic should be written first, and always expressed even though it be zero, in order to avoid error due to forgetting it.

2. *To find the mantissa from the table.* (*a*) When the number has just three significant figures. Pages 224 and 225 give a table of the mantissæ of the Briggs logarithms of all integers from 1 to 1000. In order to find the mantissa of a given number, look for the first two digits in the column marked N. These indicate the row in which the mantissa is to be found. The column is designated by the third digit. Thus the mantissa for 478 is 6794, and the entire logarithm is 2.6794 by 1.

(*b*) When the number has less than three significant digits. To find the logarithm of .7 we look for 700 in the tables and find the mantissa 8451. Hence $\log .7 = \bar{1}.8451$, or $9.8451 - 10$.

COLLEGE ALGEBRA

N	0	1	2	3	4	5	6	7	8	9
0	0000	0000	3010	4771	6021	6990	7782	8451	9031	9542
1	0000	0414	0792	1139	1461	1761	2041	2304	2553	2788
2	3010	3222	3424	3617	3802	3979	4150	4314	4472	4624
3	4771	4914	5051	5185	5315	5441	5563	5682	5798	5911
4	6021	6128	6232	6335	6435	6532	6628	6721	6812	6902
5	6990	7076	7160	7243	7324	7404	7482	7559	7634	7709
6	7782	7853	7924	7993	8062	8129	8195	8261	8325	8388
7	8451	8513	8573	8633	8692	8751	8808	8865	8921	8976
8	9031	9085	9138	9191	9243	9294	9345	9395	9445	9494
9	9542	9590	9638	9685	9731	9777	9823	9868	9912	9956
10	0000	0043	0086	0128	0170	0212	0253	0294	0334	0374
11	0414	0453	0492	0531	0569	0607	0645	0682	0719	0755
12	0792	0828	0864	0899	0934	0969	1004	1038	1072	1106
13	1139	1173	1206	1239	1271	1303	1335	1367	1399	1430
14	1461	1492	1523	1553	1584	1614	1644	1673	1703	1732
15	1761	1790	1818	1847	1875	1903	1931	1959	1987	2014
16	2041	2068	2095	2122	2148	2175	2201	2227	2253	2279
17	2304	2330	2355	2380	2405	2430	2455	2480	2504	2529
18	2553	2577	2601	2625	2648	2672	2695	2718	2742	2765
19	2788	2810	2833	2856	2878	2900	2923	2945	2967	2989
20	3010	3032	3054	3075	3096	3118	3139	3160	3181	3201
21	3222	3243	3263	3284	3304	3324	3345	3365	3385	3404
22	3424	3444	3464	3483	3502	3522	3541	3560	3579	3598
23	3617	3636	3655	3674	3692	3711	3729	3747	3766	3784
24	3802	3820	3838	3856	3874	3892	3909	3927	3945	3962
25	3979	3997	4014	4031	4048	4065	4082	4099	4116	4133
26	4150	4166	4183	4200	4216	4232	4249	4265	4281	4298
27	4314	4330	4346	4362	4378	4393	4409	4425	4440	4456
28	4472	4487	4502	4518	4533	4548	4564	4579	4594	4609
29	4624	4639	4654	4669	4683	4698	4713	4728	4742	4757
30	4771	4786	4800	4814	4829	4843	4857	4871	4886	4900
31	4914	4928	4942	4955	4969	4983	4997	5011	5024	5038
32	5051	5065	5079	5092	5105	5119	5132	5145	5159	5172
33	5185	5198	5211	5224	5237	5250	5263	5276	5289	5302
34	5315	5328	5340	5353	5366	5378	5391	5403	5416	5428
35	5441	5453	5465	5478	5490	5502	5514	5527	5539	5551
36	5563	5575	5587	5599	5611	5623	5635	5647	5658	5670
37	5682	5694	5705	5717	5729	5740	5752	5763	5775	5786
38	5798	5809	5821	5832	5843	5855	5866	5877	5888	5899
39	5911	5922	5933	5944	5955	5966	5977	5988	5999	6010
40	6021	6031	6042	6053	6064	6075	6085	6096	6107	6117
41	6128	6138	6149	6160	6170	6180	6191	6201	6212	6222
42	6232	6243	6253	6263	6274	6284	6294	6304	6314	6325
43	6335	6345	6355	6365	6375	6385	6395	6405	6415	6425
44	6435	6444	6454	6464	6474	6484	6493	6503	6513	6522
45	6532	6542	6551	6561	6571	6580	6590	6599	6609	6618
46	6628	6637	6646	6656	6665	6675	6684	6693	6702	6712
47	6721	6730	6739	6749	6758	6767	6776	6785	6794	6803
48	6812	6821	6830	6839	6848	6857	6866	6875	6884	6893
49	6902	6911	6920	6928	6937	6946	6955	6964	6972	6981
N	0	1	2	3	4	5	6	7	8	9

LOGARITHMS

N	0	1	2	3	4	5	6	7	8	9
50	6990	6998	7007	7016	7024	7033	7042	7050	7059	7067
51	7076	7084	7093	7101	7110	7118	7126	7135	7143	7152
52	7160	7168	7177	7185	7193	7202	7210	7218	7226	7235
53	7243	7251	7259	7267	7275	7284	7292	7300	7308	7316
54	7324	7332	7340	7348	7356	7364	7372	7380	7388	7396
55	7404	7412	7419	7427	7435	7443	7451	7459	7466	7474
56	7482	7490	7497	7505	7513	7520	7528	7536	7543	7551
57	7559	7566	7574	7582	7589	7597	7604	7612	7619	7627
58	7634	7642	7649	7657	7664	7672	7679	7686	7694	7701
59	7709	7716	7723	7731	7738	7745	7752	7760	7767	7774
60	7782	7789	7796	7803	7810	7818	7825	7832	7839	7846
61	7853	7860	7868	7875	7882	7889	7896	7903	7910	7917
62	7924	7931	7938	7945	7952	7959	7966	7973	7980	7987
63	7993	8000	8007	8014	8021	8028	8035	8041	8048	8055
64	8062	8069	8075	8082	8089	8096	8102	8109	8116	8122
65	8129	8136	8142	8149	8156	8162	8169	8176	8182	8189
66	8195	8202	8209	8215	8222	8228	8235	8241	8248	8254
67	8261	8267	8274	8280	8287	8293	8299	8306	8312	8319
68	8325	8331	8338	8344	8351	8357	8363	8370	8376	8382
69	8388	8395	8401	8407	8414	8420	8426	8432	8439	8445
70	8451	8457	8463	8470	8476	8482	8488	8494	8500	8506
71	8513	8519	8525	8531	8537	8543	8549	8555	8561	8567
72	8573	8579	8585	8591	8597	8603	8609	8615	8621	8627
73	8633	8639	8645	8651	8657	8663	8669	8675	8681	8686
74	8692	8698	8704	8710	8716	8722	8727	8733	8739	8745
75	8751	8756	8762	8768	8774	8779	8785	8791	8797	8802
76	8808	8814	8820	8825	8831	8837	8842	8848	8854	8859
77	8865	8871	8876	8882	8887	8893	8899	8904	8910	8915
78	8921	8927	8932	8938	8943	8949	8954	8960	8965	8971
79	8976	8982	8987	8993	8998	9004	9009	9015	9020	9025
80	9031	9036	9042	9047	9053	9058	9063	9069	9074	9079
81	9085	9090	9096	9101	9106	9112	9117	9122	9128	9133
82	9138	9143	9149	9154	9159	9165	9170	9175	9180	9186
83	9191	9196	9201	9206	9212	9217	9222	9227	9232	9238
84	9243	9248	9253	9258	9263	9269	9274	9279	9284	9289
85	9294	9299	9304	9309	9315	9320	9325	9330	9335	9340
86	9345	9350	9355	9360	9365	9370	9375	9380	9385	9390
87	9395	9400	9405	9410	9415	9420	9425	9430	9435	9440
88	9445	9450	9455	9460	9465	9469	9474	9479	9484	9489
89	9494	9499	9504	9509	9513	9518	9523	9528	9533	9538
90	9542	9547	9552	9557	9562	9566	9571	9576	9581	9586
91	9590	9595	9600	9605	9609	9614	9619	9624	9628	9633
92	9638	9643	9647	9652	9657	9661	9666	9671	9675	9680
93	9685	9689	9694	9699	9703	9708	9713	9717	9722	9727
94	9731	9736	9741	9745	9750	9754	9759	9763	9768	9773
95	9777	9782	9786	9791	9795	9800	9805	9809	9814	9818
96	9823	9827	9832	9836	9841	9845	9850	9854	9859	9863
97	9868	9872	9877	9881	9886	9890	9894	9899	9903	9908
98	9912	9917	9921	9926	9930	9934	9939	9943	9948	9952
99	9956	9961	9965	9969	9974	9978	9983	9987	9991	9996
N	0	1	2	3	4	5	6	7	8	9

COLLEGE ALGEBRA

(c) When the number has more than three significant digits.

Thus in the case of log 32.456, since 32.456 lies .56 of the way from 32.4 to 32.5, its logarithm must lie about .56 of the way from log 32.4 to log 32.5. But log $32.4 = 1.5105$ and log $32.5 = 1.5119$. Hence, leaving the decimal point out of account, the increase, or *tabular difference* as it is called, is 14, and .56 of this is 7.84. Hence, adding this correction to log 32.4, we have log $32.456 = 1.5113$. Since our tables are given only to four decimal places, we retain only four in correction, always figuring the fourth place to the nearest unit, thus in this case adding 8 as the correction.

Likewise for log .0035678, we find from the tables log $.00356 = 7.5514 - 10$, and the tabular difference is 13. The correction $(13 \times .78)$ is 10, so that log $.0035678 = 7.5524 - 10$.

Find the logarithms of the following numbers:

1. 428.
2. 327.
3. 82.46.
4. 32.875.
5. .524.
6. .02345.
7. .38634.
8. 430.23.
9. .050009.
10. .0042085.
11. 20.308.
12. 7,352,000.

277. To find the number corresponding given logarithm.
Given log $x = 3.3765$; to find x.

This is the reverse operation of that given in **276**. Since the characteristic merely determines the position of the decimal point, **273**, we look for the mantissa in the table. The next smaller mantissa in the table is 3747, which corresponds to the number 237. The *excess* of 3765 over 3747 is 18. The tabular difference as found from the table by subtracting 3747 from 3766 is 19. Hence 3765 is $\frac{18}{19}$ of the way from 3747 to 3766, and the corresponding number is about $\frac{18}{19}$ of the way from 237 to 238, or, reducing $\frac{18}{19}$ to a decimal,

LOGARITHMS

about .95 of a unit beyond 237. Hence the corresponding digits are 23795, and since the characteristic is 3, the decimal point must be moved three places to the right of where it would be placed for a zero characteristic, and hence $x = 2379.5$.

Similarly, if $\log y = 7.3765 - 10$, $y = .0023795$, since in this case the characteristic is $7 - 10$, or -3, and the decimal point must be moved three places to the left.

Find the numbers corresponding to the following logarithms:

1. 2.8987.
2. 3.5705.
3. 0.7016.
4. 9.8814 − 10.
5. 1.9542.
6. 3.6558.
7. 4.6062.
8. 2.6842.
9. 1.3427.
10. 0.4850.
11. 7.6123.
12. 8.5493 − 10.
13. 5.8124.
14. 0.7318.
15. 4.7306 − 10.
16. $\bar{5}.4783$.
17. $\bar{2}.7005$.
18. $\bar{1}.6100$.

278. Cologarithms. *The* **cologarithm** *of a number is the logarithm of the reciprocal of the number.*

Thus $\operatorname{colog} 425 = \log \dfrac{1}{425} = \log 1 - \log 425$ **(269)**

$$= 0 - 2.6284.$$

But since we always wish to have the mantissa of a logarithm positive, we write $0 = 10 - 10$, and subtract 2.6284 from this, as follows:

$$\log 1 = 10.0000 - 10$$
$$\log 425 = 2.6284$$
$$\overline{\operatorname{colog} 425 = 7.3716 - 10.}$$

COLLEGE ALGEBRA

In practice this is done mentally by beginning at the left and subtracting each digit from 9, except the last significant digit, which is subtracted from 10.

279. Computation by logarithms.

EXAMPLES

1. Find the value of $2345 \times 2.327 \times .004296$.

$\log (2345 \times 2.327 \times .004296) = \log 2345 + \log 2.327$
$\qquad\qquad\qquad\qquad\qquad\qquad + \log .004296.$ **(269)**

$\log 2345 = 3.3702$
$\log 2.327 = 0.3668$
$\log .004296 = 7.6331 - 10$
$\overline{\log (2345 \times 2.327 \times .004296) = 1.3701.}$

$\qquad\qquad\qquad\qquad\quad \dfrac{3692}{19)\overline{9.00}}$
$\qquad\qquad\qquad\qquad\qquad .47$

Therefore, $2345 \times 2.327 \times .004296 = 23.447$.

NOTE. It must be borne in mind that these logarithms are only approximations carried out to a number of decimal places determined by the number of decimal places of the table used. Four-place tables do not give even five figures in the result with any considerable degree of accuracy. If greater accuracy is desired, a table with more decimal places must be used.

2. Find the value of $\dfrac{37.54 \times .02436}{.002578 \times 386.5}$.

Since this is the product of two numbers divided by the product of two others, it might be solved by subtracting the sum of the logarithms of the factors of the denominator from the sum of the logarithms of the factors of the numerator. But this would require several operations, and it is customary to regard the fraction as the continued product of the factors of the numerator and the reciprocals of the factors of the denominator.

LOGARITHMS

Thus
$$\log \frac{37.54 \times .02436}{.002578 \times 386.5} = \log\left(37.54 \times .02436 \times \frac{1}{.002578} \times \frac{1}{386.5}\right)$$
$$= \log 37.54 + \log .02436 + \operatorname{colog} .002578 + \operatorname{colog} 386.5.$$

$$\log 37.54 = 1.5745$$
$$\log .02436 = 8.3867 - 10$$
$$\operatorname{colog} .002578 = 2.5887$$
$$\operatorname{colog} 386.5 = 7.4128 - 10$$
$$\log \frac{37.54 \times .02436}{.002578 \times 386.5} = 9.9627 - 10.$$

Hence $\dfrac{37.54 \times .02436}{.002578 \times 386.5} = .91775.$

3. Find the value of $(38.64)^6$.

$$\log (38.64)^6 = 6 \times \log 38.64. \quad (\mathbf{269})$$
$$\log 38.64 = 1.5870.$$

Therefore $\log (38.64)^6 = 9.5220,$
and $(38.64)^6 = 3{,}326{,}900{,}000.$

4. Find the value of $\sqrt[5]{.7684}$.

$$\log \sqrt[5]{.7684} = \tfrac{1}{5} \log .7684.$$
$$\log .7684 = 9.8856 - 10$$
$$= 49.8856 - 50.$$

Therefore $\log \sqrt[5]{.7684} = 9.9771 - 10,$
and $\sqrt[5]{.7684} = .9486.$

COLLEGE ALGEBRA

Find by means of logarithms the approximate values of the following expressions:

5. $234 \times 345 \times 456$.

6. $2345 \times 3456 \times 4567$.

7. $86.23 \times 23.45 \times .08632$.

8. $\dfrac{28.42 \times 67.54 \times 96.72}{92.57 \times 13.83}$.

9. $\dfrac{48.87 \times .03245 \times 389.2}{7.459 \times 9.351}$.

10. $\dfrac{.07371 \times 907.3 \times .6007}{.7623 \times 8.076}$.

11. $\dfrac{-7.831 \times 3.867 \times 8.903}{.007459 \times 12.87}$.

NOTE. Since all real powers of 10 are positive, negative numbers have no real logarithms. Hence examples involving negative numbers must be worked as if the numbers were positive, and then the proper sign is to be attached to result.

12. $\dfrac{(-.008734) \times (-8.345) \times 834.7}{(-.8793) \times (-900.6)}$.

13. $(8.341)^4$.

14. $(.05376)^5$.

15. $(38.56)^{\frac{1}{2}}$.

16. $(56.38)^{\frac{1}{4}}$.

17. $(.02583)^{\frac{1}{5}}$.

18. $(-8.425)^{\frac{1}{3}}$.

19. $(-.5429)^{\frac{2}{3}}$.

20. $(-.05387)^{\frac{1}{4}}$.

21. $(.384\frac{4}{7})^{\frac{1}{4}}$.

22. $\dfrac{\sqrt{3596} \times \sqrt[3]{.4287}}{\sqrt[4]{.0586}}$.

23. $\sqrt{\sqrt[6]{.5804} \times \sqrt[3]{.2405}}$.

24. Find by means of logarithms the amount of $486 in five years at five per cent if the interest is compounded annually.

25. Find the amount of $384 in forty years at four per cent if the interest is compounded semiannually.

26. Solve the equation $4^x = 246$.

LOGARITHMS

Taking the logarithm of each member, we have

$$x \log 4 = \log 246,$$

or $$x = \frac{\log 246}{\log 4} = \frac{2.3909}{0.6021} = 3.971.$$

27. Solve the equation $415^{2x+3} = 517^{12x-1}$.

28. Find the logarithm of 428 to the base .8.

If x represent the required logarithm,

$$.8^x = 428,$$

and $$x = \frac{\log 428}{\log .8} \quad \text{(Compare with 271)}$$

$$= \frac{2.6314}{9.9031 - 10} = -\frac{2.6314}{0.0969} = -27.16.$$

NOTE. In this case $9.9031 - 10$ is really a binomial expression and hence must be combined in a single monomial before ordinary arithmetical division can be performed.

29. Find the logarithm of 376 to the base 12.

30. Find the logarithm of .536 to the base 7.

THE EXPONENTIAL FUNCTION

280. Definition. *Let us define $F(x)$ as the limit of $\left(1 + \dfrac{x}{n}\right)^n$, as $n \doteq \infty$, that is, $F(x) = \underset{n \doteq \infty}{L} \left(1 + \dfrac{x}{n}\right)^n$, for both real and complex values of x and n.*

281. From the foregoing definition we easily derive the following properties of $F(x)$.

1. When $x = 0$, we have $F(0) = L(1)^n \doteq 1.$*

---

* Although 1^∞ must be considered an indeterminate form when it is the limit of a variable which approaches 1 raised to an infinite power, here we have strictly a constant.

2. For all real values of x and n, $F(x)$ is the same for a given value of x in whatever way n becomes indefinitely great. For if n is any real positive fraction, we can always choose two consecutive positive integers such that

$$m+1 > n > m.$$

Hence, according as x is positive or negative,

$$\frac{x}{m+1} \lessgtr \frac{x}{n} \lessgtr \frac{x}{m},$$

and
$$1 + \frac{x}{m} \gtrless 1 + \frac{x}{n} \gtrless 1 + \frac{x}{m+1},$$

and hence
$$\left(1+\frac{x}{m}\right)^{m+1} > \left(1+\frac{x}{n}\right)^n > \left(1+\frac{x}{m+1}\right)^m,$$

or
$$\left(1+\frac{x}{m}\right)\left(1+\frac{x}{m}\right)^m > \left(1+\frac{x}{n}\right)^n > \frac{\left(1+\frac{x}{m+1}\right)^{m+1}}{\left(1+\frac{x}{m+1}\right)} \quad \text{for } x > 0\text{*}.$$

Taking the limits as $n \doteq \infty$, and hence as $m \doteq \infty$, and $m+1 \doteq \infty$, and denoting by $f(x)$ the value of the limit of $\left(1+\frac{x}{m}\right)^m$ as m, while remaining a positive integer, becomes indefinitely great, we see that $\left(1+\frac{x}{n}\right)^n$, lying between two numbers each of which approaches the limit $f(x)$, must itself approach $f(x)$ as its limit, or

$$\underset{n \doteq \infty}{L}\left(1+\frac{x}{n}\right)^n = f(x),$$

when n is positive, whether x is positive or negative.

* For $x < 0$, the inequality sign and the order of the exponents would be reversed.

LOGARITHMS

If $n = -p$, where p is a positive number, then

$$\mathop{L}_{n \pm \infty}\left(1 + \frac{x}{n}\right)^n = \mathop{L}_{p \pm \infty}\left(1 - \frac{x}{p}\right)^{-p} = \mathop{L}_{p \pm \infty}\left(\frac{p}{p-x}\right)^p$$

$$= \mathop{L}_{p-x \pm \infty}\left(1 + \frac{x}{p-x}\right)^{p-x}\left(1 + \frac{x}{p-x}\right)^x = f(x).$$

We have proved therefore that

$$\mathop{L}_{n \pm \infty}\left(1 + \frac{x}{n}\right)^n = f(x)$$

for all real values of x and n.

If $x = 1$, we have $\mathop{L}_{n \pm \infty}\left(1 + \frac{1}{n}\right)^n = f(1)$.

3. Since
$$\mathop{L}_{n \pm \infty}\left(1 + \frac{x}{n}\right)^n = \mathop{L}_{n \pm \infty}\left(\left(1 + \frac{x}{n}\right)^{\frac{n}{x}}\right)^x,$$

we have, putting $\dfrac{n}{x} = m$,

$$\mathop{L}_{n \pm \infty}\left(1 + \frac{x}{n}\right)^n = \mathop{L}_{m \pm \infty}\left(\left(1 + \frac{1}{m}\right)^m\right)^x = f(1)^x,$$

or $f(x) = f(1)^x$, for all real values of x.

In particular, $\qquad f(-x) = f(1)^{-x}.$

4. By (3), we have
$$f(x)f(y) = f(1)^x f(1)^y = f(1)^{x+y} = f(x+y),$$
or $\qquad f(x)f(y) = f(x+y).$

The second property shows that for all real values of x and n,
$$F(x) = f(x).$$

282. Expanding $\left(1+\dfrac{x}{n}\right)^n$ by the binomial theorem, we have

$$\left(1+\frac{x}{n}\right)^n = 1 + \frac{n}{n}x + \frac{n(n-1)}{2!}\frac{x^2}{n^2} + \cdots$$

$$+ \frac{n(n-1)\cdots(n-r+1)}{r!}\frac{x^r}{n^r} + \cdots = 1 + x + \frac{\left(1-\dfrac{1}{n}\right)}{2!}x^2 + \cdots$$

$$+ \frac{\left(1-\dfrac{1}{n}\right)\left(1-\dfrac{2}{n}\right)\cdots\left(1-\dfrac{r-1}{n}\right)}{r!}x^r + \cdots. \quad (1)$$

Taking the limits of both sides as $n \doteq \infty$, we have

$$f(x) = 1 + x + \frac{x^2}{2!} + \frac{x^3}{3!} + \cdots.$$

That the right-hand member has this limit may be seen as follows:

The $(r+1)$th term is

$$t_{r+1} = \frac{\left(1-\dfrac{1}{n}\right)\left(1-\dfrac{2}{n}\right)\cdots\left(1-\dfrac{r-1}{n}\right)}{r!}x^r.$$

Let $a_1, a_2, a_3, \cdots$, be positive proper fractions. Then, since

$$(1-a_1)(1-a_2) = 1 - a_1 - a_2 + a_1 a_2,$$

$$1 > (1-a_1)(1-a_2) > 1 - a_1 - a_2.$$

Similarly

$$1 > (1-a_1)(1-a_2)(1-a_3) > 1 - a_1 - a_2 - a_3,$$

. . . .

$$1 > (1-a_1)(1-a_2)\cdots(1-a_k) > 1 - a_1 - a_2 - a_3 \cdots - a_k. \quad (2)$$

LOGARITHMS

Choosing $k = r-1$, $a_1 = \dfrac{1}{n}$, $a_2 = \dfrac{2}{n}$, $\cdots$, $a_k = \dfrac{r-1}{n}$, and multiplying (2) by $\dfrac{x^r}{r!}$, we have, using a positive x for convenience,

$$\frac{x^r}{r!} < t_{r+1} > \frac{1 - \left(\dfrac{1}{n} + \dfrac{2}{n} + \cdots + \dfrac{r-1}{n}\right)}{r!} x^r,$$

or
$$\frac{x^r}{r!} > t_{r+1} > \frac{x^r}{r!} - \frac{\tfrac{1}{2}\, r(r-1)}{n\; r!} x^r,$$

$$\frac{x^r}{r!} > t_{r+1} > \frac{x^r}{r!} - \frac{x^r}{2n\,(r-2)!}.$$

Giving r the values $0, 2, 4, \cdots$, and the values $1, 3, 5, \cdots$, we get the following two systems of relations respectively,

$$1 = t_1 = 1, \qquad\qquad x = t_2 = x,$$
$$\frac{x^2}{2!} > t_3 > \frac{x^2}{2!} - \frac{x^2}{2n}, \qquad \frac{x^3}{3!} > t_4 > \frac{x^3}{3!} - \frac{x^3}{2n}, \qquad (3)$$
$$\frac{x^4}{4!} > t_5 > \frac{x^4}{4!} - \frac{x^4}{2n\,2!}, \qquad \frac{x^5}{5!} > t_6 > \frac{x^5}{5!} - \frac{x^5}{2n\,3!},$$
$$\cdots\cdots\cdots\cdots \qquad\qquad \cdots\cdots\cdots\cdots$$

Adding the corresponding terms of the relations of each system, we have

$$1 + \frac{x^2}{2!} + \frac{x^4}{4!} + \cdots > t_1 + t_3 + t_5 + \cdots > 1 + \frac{x^2}{2!} + \frac{x^4}{4!} + \cdots$$
$$- \frac{x^2}{2n}\left(1 + \frac{x^2}{2!} + \frac{x^4}{4!} + \cdots\right),$$

$$x + \frac{x^3}{3!} + \frac{x^5}{5!} + \cdots > t_2 + t_4 + t_6 + \cdots > x + \frac{x^3}{3!} + \frac{x^5}{5!} + \cdots$$
$$- \frac{x^2}{2n}\left(x + \frac{x^3}{3!} + \frac{x^5}{5!} + \cdots\right).$$

As $n \doteq \infty$, the series $1 + \dfrac{x^2}{2!} + \dfrac{x^4}{4!} + \cdots$

and $\qquad x + \dfrac{x^3}{3!} + \dfrac{x^5}{5!} + \cdots$

are absolutely convergent, **212**, 16, the terms

$$\frac{x^2}{2n}\left(1 + \frac{x^2}{2!} + \frac{x^4}{4!} + \cdots\right)$$

and $\qquad \dfrac{x^2}{2n}\left(x + \dfrac{x^3}{3!} + \dfrac{x^5}{5!} + \cdots\right)$

approach zero, and hence we see that

$$\underset{n \doteq \infty}{L}(t_1 + t_3 + t_5 + \cdots) = 1 + \frac{x^2}{2!} + \frac{x^4}{4!} + \cdots,$$

$$\underset{n \doteq \infty}{L}(t_2 + t_4 + t_6 + \cdots) = x + \frac{x^3}{3!} + \frac{x^5}{5!} + \cdots.$$

Adding and subtracting these limits and remembering that

$$t_1 + t_2 + \cdots + t_{n+1} = \left(1 + \frac{x}{n}\right)^n,$$

$$t_1 - t_2 + \cdots + (-1)^n t_{n+1} = \left(1 - \frac{x}{n}\right)^n,$$

we have $\qquad \underset{n \doteq \infty}{L}\left(1 + \dfrac{x}{n}\right)^n = 1 + x + \dfrac{x^2}{2!} + \dfrac{x^3}{3!} + \cdots,$ **(213)**

$$\underset{n \doteq \infty}{L}\left(1 - \frac{x}{n}\right)^n = 1 - x + \frac{x^2}{2!} - \frac{x^3}{3!} + \cdots.$$

LOGARITHMS

This completes the proof that for any real values of x and n,

$$f(x) = \operatorname*{L}_{n\to\infty}\left(1+\frac{x}{n}\right)^n = 1 + x + \frac{x^2}{2!} + \frac{x^3}{3!} + \cdots.$$

If in the preceding proof we had taken x negative, the right-hand column of relations (3.) would have had their signs all reversed, but in taking the limits we would have arrived at the same result.*

283. The series

$$f(1) = 1 + 1 + \frac{1}{2!} + \frac{1}{3!} + \cdots$$

is denoted by e, which is the base of the natural system.

That the quantity e is finite and lies between 2 and 3 may be easily seen, for

$$e = 1 + 1 + \frac{1}{2!} +$$
$$= 2+$$

and since $\quad 1 + \frac{1}{2!} + \frac{1}{3!} + \cdots < 1 + \frac{1}{2} + \frac{1}{2^2} + \cdots,$

or $\qquad\qquad\qquad e - 1 < \dfrac{1}{1-\frac{1}{2}},$

or $\qquad\qquad\qquad e - 1 < 2.$

Therefore $e < 3$, hence $3 > e > 2$.

Since $f(x) = f(1)^x$, we have proved that

$$e^x = 1 + x + \frac{x^2}{2!} + \frac{x^3}{3!} + \cdots$$

for all real values of x.

* Or otherwise $\quad \operatorname*{L}_{n\to\infty}\left(1-\dfrac{x}{n}\right)^n = \operatorname*{L}_{n\to\infty}\left\{\left(1+\dfrac{x}{-n}\right)^{-n}\right\}^{-1}$

$= \{f(x)\}^{-1} = \dfrac{1}{f(x)} = \dfrac{1}{f(1)^x} = \{f(1)\}^{-x} = f(-x).$

284. The determination of the value of $F(x)$ when x and n are complex numbers can be made as follows:

If z is a complex number and n a positive integer, the $(r+1)$th term in $\left(1+\dfrac{z}{n}\right)^n$, z being equal to $x(\cos\phi + i\sin\phi)$, is, **380**,

$$T_{r+1} = \frac{\left(1-\dfrac{1}{n}\right)\left(1-\dfrac{2}{n}\right)\cdots\left(1-\dfrac{r-1}{n}\right)x^r(\cos r\phi + i\sin r\phi)}{r!}.$$

Let $\quad t'_{r+1} = \dfrac{\left(1-\dfrac{1}{n}\right)\left(1-\dfrac{2}{n}\right)\cdots\left(1-\dfrac{r-1}{n}\right)x^r\cos r\phi}{r!},$

and $\quad t''_{r+1} = \dfrac{\left(1-\dfrac{1}{n}\right)\left(1-\dfrac{2}{n}\right)\cdots\left(1-\dfrac{r-1}{n}\right)x^r\sin r\phi}{r!},$

then as before $\quad 1 = t'_1 = 1,$

$$x\cos\phi = t'_2 = x\cos\phi,$$

$$\frac{x^2}{2!}\cos 2\phi > t'_3 > \frac{x^2}{2!}\cos 2\phi - \frac{x^2\cos 2\phi}{2n},$$

$$\frac{x^3}{3!}\cos 3\phi > t'_4 > \frac{x^3}{3!}\cos 3\phi - \frac{x^3\cos 3\phi}{2n},$$

$$\frac{x^4}{4!}\cos 4\phi > t'_5 > \frac{x^4}{4!}\cos 4\phi - \frac{x^4\cos 4\phi}{2n\,2!},$$

$$\frac{x^5}{5!}\cos 5\phi > t'_6 > \frac{x^5}{5!}\cos 5\phi - \frac{x^5\cos 5\phi}{2n\,3!},$$

. . . .

$$\frac{x^n}{n!}\cos n\phi > t'_{n+1} > \frac{x^n}{n!}\cos n\phi - \frac{x^n\cos n\phi}{2n\,(n-2)!}.$$

LOGARITHMS

On adding, we get

$$1 + x\cos\phi + \frac{x^2}{2!}\cos 2\phi + \cdots + \frac{x^n}{n!}\cos n\phi > t'_1 + t'_2 + \cdots + t'_{n+1}$$

$$> 1 + x\cos\phi + \cdots + \frac{x^n}{n!}\cos n\phi$$

$$- \frac{x^2}{2n}\left(\cos 2\phi + x\cos 3\phi + \cdots + \frac{x^{n-2}}{(n-2)!}\cos n\phi\right).$$

As $n \doteq \infty$, the series

$$1 + x\cos\phi + \cdots + \frac{x^n}{n!}\cos n\phi$$

and

$$\cos 2\phi + x\cos 3\phi + \cdots + \frac{x^{n-2}}{(n-2)!}\cos n\phi$$

are absolutely convergent, since the terms of each are less than the corresponding terms of the absolutely convergent series

$$1 + x + \frac{x^2}{2!} + \cdots + \frac{x^n}{n!},$$

which as $n \doteq \infty$ has the limit $f(x)$ or e^x. Therefore

$$\underset{n\doteq\infty}{L}(t'_1 + t'_2 + \cdots + t'_{n+1}) = 1 + x\cos\phi + \frac{x^2}{2!}\cos 2\phi + \cdots. \quad (1)$$

In the same way it can be proved that

$$\underset{n\doteq\infty}{L}(t''_2 + t''_3 + \cdots + t''_{n+1}) = x\sin\phi + \frac{x^2}{2!}\sin 2\phi + \cdots. \quad (2)$$

Multiplying (2) by i and adding the result to (1) and observing that

$$t'_1 = T_1,\ t'_2 + it''_2 = T_2,\ t'_3 + it''_3 = T_3,\ \text{etc.},$$

and that $\quad (T_1 + T_2 + \cdots + T_{n+1}) = \left(1 + \frac{z}{n}\right)^n, \quad$ **(213, 214)** we have

$$\underset{n\doteq\infty}{L}(T_1+T_2+\cdots+T_{n+1})=\underset{n\doteq\infty}{L}\left(1+\frac{z}{n}\right)^n=1+x(\cos\phi+i\sin\phi)$$
$$+\frac{x^2}{2!}(\cos 2\phi+i\sin 2\phi)+\cdots,$$

or
$$F(z)=1+z+\frac{z^2}{2!}+\frac{z^3}{3!}+\cdots.$$

The extension to any real value of n may be made by considering the limits of the moduli and arguments of $\left(1+\frac{z}{m+1}\right)^m, \left(1+\frac{z}{n}\right)^n, \left(1+\frac{z}{m}\right)^{m+1}$ and finally extending to a negative n.

Hence, when z is complex and n is real,

$$F(z)=f(z).$$

If both z and n are complex, we have, if
$n=m(\cos\phi+i\sin\phi),$

$$\underset{n\doteq\infty}{L}\left(1+\frac{z}{n}\right)^n=\underset{m\doteq\infty}{L}\left(1+\frac{z}{m(\cos\phi+i\sin\phi)}\right)^{m(\cos\phi+i\sin\phi)}$$
$$=\underset{m\doteq\infty}{L}\left(\left(1+\frac{z(\cos\phi-i\sin\phi)}{m}\right)^m\right)^{\cos\phi+i\sin\phi}$$
$$=(f(z(\cos\phi-i\sin\phi)))^{\cos\phi+i\sin\phi}.$$

This is as far as we can carry the proof. If, however, we agree to give to a complex exponent such an interpretation that the third property, viz. $f(z)=(f(1))^z$, shall still hold even when z is complex, we have

$$F(z)=(f(z(\cos\phi-i\sin\phi)))^{\cos\phi+i\sin\phi}$$
$$=f(1)^z$$
$$=f(z).$$

LOGARITHMS

Thus, for all values of x and n, we have
$$F(x) = f(x).$$

That the series denoted by $f(x)$ is convergent has been seen from the mode of its derivation, since each of the constituent series of which it is composed is convergent whether x be real or complex. The result
$$\underset{n \doteq \infty}{L} \left(1 + \frac{x}{n}\right)^n = e^x = 1 + x + \frac{x^2}{2!} + \frac{x^3}{3!} + \cdots$$
is known as the *Exponential Theorem*.

285. Since $a^x = e^{x \log_e a}$, we have
$$a^x = 1 + (\log_e a) x + (\log_e a)^2 \frac{x^2}{2!} + \cdots. \tag{1}$$

Substituting in this $1+y$ for a, we have
$$(1+y)^x = 1 + x \log_e(1+y) + \frac{x^2}{2!}(\log_e(1+y))^2 + \cdots. \tag{2}$$

If y is numerically less than unity, we can expand by the binomial theorem and get
$$1 + xy + \frac{x(x-1)}{2!} y^2 + \frac{x(x-1)(x-2)}{3!} y^3 + \cdots$$
$$= 1 + x \log_e(1+y) + \frac{x^2}{2!}(\log_e(1+y))^2 + \cdots. \tag{3}$$

Equating the coefficients of x on both sides of (3), we have (**212**, 17)
$$\log_e(1+y) = y - \frac{y^2}{2} + \frac{y^3}{3} - \frac{y^4}{4} + \cdots. \tag{4}$$

This is called the *logarithmic series*.

Changing y into $-y$, we have
$$\log_e(1-y) = -y - \frac{y^2}{2} - \frac{y^3}{3} - \frac{y^4}{4} - \cdots. \tag{5}$$

The logarithmic series may be used to find the logarithm of any number, but since the series converges so slowly, it is more expedient to use the following:

Subtracting the corresponding members of (4) and (5), we have

$$\log_e(1+y) - \log_e(1-y) = \left(y - \frac{y^2}{2} + \frac{y^3}{3} - \cdots\right)$$
$$- \left(-y - \frac{y^2}{2} - \frac{y^3}{3} - \cdots\right),$$

or
$$\log_e \frac{1+y}{1-y} = 2\left(y + \frac{y^3}{3} + \frac{y^5}{5} + \cdots\right). \quad (6)$$

Substitute in this $\frac{1+y}{1-y} = \frac{m}{n}$, that is, $y = \frac{m-n}{m+n}$, and it becomes

$$\log_e \frac{m}{n} = 2\left(\frac{m-n}{m+n} + \frac{1}{3}\left(\frac{m-n}{m+n}\right)^3 + \frac{1}{5}\left(\frac{m-n}{m+n}\right)^5 + \cdots\right). \quad (7)$$

If $m = n$, $\log_e 1 = 0$.

If $m = 2$ and $n = 1$, $\log_e 2 = 2\left(\frac{1}{3} + \frac{1}{3}\left(\frac{1}{3}\right)^3 + \frac{1}{5}\left(\frac{1}{3}\right)^5 + \cdots\right).$

If $m = n+1$, (7) becomes

$$\log_e \frac{n+1}{n} = 2\left(\frac{1}{2n+1} + \frac{1}{3}\left(\frac{1}{2n+1}\right)^3 + \frac{1}{5}\left(\frac{1}{2n+1}\right)^5 + \cdots\right),$$

and this is equivalent to

$$\log_e(n+1) = \log_e n + 2\left(\frac{1}{2n+1} + \frac{1}{3}\left(\frac{1}{2n+1}\right)^3 + \frac{1}{5}\left(\frac{1}{2n+1}\right)^5 + \cdots\right). \quad (8)$$

LOGARITHMS

From (8) the logarithms of all numbers to the base e may be obtained. It has been shown in **271** how to change from one base to another.

To obtain the logarithms of numbers to the base 10 we have to multiply the logarithms of the numbers to the base e by $\dfrac{1}{\log_e 10}$, which may be found from (8) to be 0.434294^+.

EXAMPLES. Compute $\log_e 2$, $\log_e 3$, $\log_e 4$, $\log_{10} 2$, $\log_{10} 3$, $\log_{10} 4$.

LOGARITHMS.

(210.) *Logarithms* are numbers, by the aid of which many arithmetical operations are greatly simplified.

In the following relations:

$$\left.\begin{array}{ll} a^x = b & (1) \\ a^y = c & (2) \\ a^z = d & (3) \\ \&c., & \end{array}\right\} \quad (A)$$

x, y, and z, are respectively the logarithms of b, c, and d.

(211.) The assumed root a is called the base of the system of logarithms.

(212.) If in (1), of equations (A), we make $x=0$, we shall have $a^0 = b = 1$, (Art. 41,) for all values of a, *therefore the logarithm of 1 is always 0.*

(213.) If in (1), we suppose the base to be negative, we shall have $(-a)^x = b$.

If b is positive, then x must be even, if b is negative, then x must be odd; hence we can not represent all values of b by the expression $(-a)^x$. *Therefore the base of every system of logarithms must be positive.*

(214.) If in (1), we suppose b to be negative, we shall have $a^x = -b$. Now, since a is always positive, the expression a^x is positive for all values of x either positive or negative.

Therefore, the logarithm of a negative quantity is impossible.

(215.) Each different base must produce a different system of logarithms; the logarithms in common use have 10 for their base.

So that we have

$10^0 = 1$; $10^1 = 10$; $10^2 = 100$; $10^3 = 1000$; &c.

Hence, we have

log. $1 = 0$
log. $10 = 1$
log. $100 = 2$
log. $1000 = 3$
log. $10000 = 4$
&c.

log. $\dfrac{1}{10} = -1$

log. $\dfrac{1}{100} = -2$

log. $\dfrac{1}{1000} = -3$

log. $\dfrac{1}{10000} = -4$

&c.

(216.) If we take the product of equations (1) and (2) of group (A), we shall have

$$a^{x+y} = bc, \qquad (4)$$

from which we discover that, *the logarithm of the product of two quantities is equal to the sum of the logarithms.*

And in general, *the logarithm of a number consisting of any number of factors is equal to the sum of the logarithms of all its factors.*

(217.) It also follows from the above, *that n times the logarithm of any number is equal to the logarithm of its nth power.*

LOGARITHMS.

(218.) If we divide equation (1) by (2), of group (A), we shall find

$$a^{x-y}=\frac{b}{c}, \qquad (5)$$

from which we see that, *the difference of the logarithms of any two quantities is equal to the logarithm of their quotient.*

(219.) From which it also follows, that, *the nth part of the logarithm of any number is equal to the logarithm of its nth root.*

(220.) We will now show how the numerical values of logarithms may be found.

If x is the logarithm of N for the base a we shall have this condition

$$a^x = N. \qquad (6)$$

If we assume

$$\left. \begin{array}{l} a = 1+m \\ N = 1+n, \end{array} \right\} \qquad (7)$$

we shall have

$$(1+m)^x = 1+n. \qquad (8)$$

Involving both members of this to the yth power, we shall have

$$(1+m)^{xy} = (1+n)^y. \qquad (9)$$

By the *Binomial Theorem*, we find

$$(1+m)^{xy} = 1 + xym + \frac{xy(xy-1)}{1.2}\cdot m^2 + \frac{xy(xy-1)(xy-2)}{1.2.3}\cdot m^3 + \&c.$$

$$(1+n)^y = 1 + yn + \frac{y(y-1)}{1.2}\cdot n^2 + \frac{y(y-1)(y-2)}{1.2.3}\cdot n^3 + \&c.$$

Equating these expanded values, rejecting the units of both expressions, we have, after dividing through by y,

$$x\left\{m+\frac{xy-1}{2}\cdot m^2+\frac{(xy-1)(xy-2)}{2.3}\cdot m^3+\&c.\right\}=$$
$$n+\frac{y-1}{2}\cdot n^2+\frac{(y-1)(y-2)}{2.3}\cdot n^3+\&c.$$

This must be true for all values of y.

When $y=0$, it becomes
$$x\{m-\tfrac{1}{2}m^2+\tfrac{1}{3}m^3-\tfrac{1}{4}m^4+\&c.\}=n-\tfrac{1}{2}n^2+\tfrac{1}{3}n^3-\tfrac{1}{4}n^4+\&c.$$

Hence,
$$x=\log. N=\log.(1+n)=\frac{n-\tfrac{1}{2}n^2+\tfrac{1}{3}n^3-\tfrac{1}{4}n^4+\&c.}{m-\tfrac{1}{2}m^2+\tfrac{1}{3}m^3-\tfrac{1}{4}m^4+\&c.} \quad (10)$$

Resubstituting $a-1$ for m, and we have
$$\log.(1+n)=\frac{n-\tfrac{1}{2}n^2+\tfrac{1}{3}n^3-\tfrac{1}{4}n^4+\&c.}{(a-1)-\tfrac{1}{2}(a-1)^2+\tfrac{1}{3}(a-1)^3-\tfrac{1}{4}(a-1)^4+\&c.} \quad (11)$$

If we assume
$$M=\frac{1}{(a-1)-\tfrac{1}{2}(a-1)^2+\tfrac{1}{3}(a-1)^3-\tfrac{1}{4}(a-1)^4+\&c.},$$
we shall have
$$\log.(1+n)=M\{n-\tfrac{1}{2}n^2+\tfrac{1}{3}n^3-\tfrac{1}{4}n^4+\tfrac{1}{5}n^5-\&c.\}. \quad (B)$$

If the base be so chosen as to render $M=1$, then formula (B) will become
$$\log.(1+n)=n-\tfrac{1}{2}n^2+\tfrac{1}{3}n^3-\tfrac{1}{4}n^4+\tfrac{1}{5}n^5-\&c. \quad (C)$$

(221.) The logarithms obtained by formula (C) are called *hyperbolic*, or *Napierian*, whilst the common logarithms given by formula (B) are called *Briggean*.

(222.) We shall hereafter denote the Napierian logarithms by the abbreviation N log., whilst the common or Briggean logarithms will be represented simply by log. Hence formula (C) will become
$$N\log.(1+n)=n-\tfrac{1}{2}n^2+\tfrac{1}{3}n^3-\tfrac{1}{4}n^4+\tfrac{1}{5}n^5-\&c. \quad (C')$$

LOGARITHMS.

(223.) By comparing formulas (B) and (C) we discover this relation

$$M \times N \log.(1+n) = \log.(1+n). \qquad (12)$$

Therefore

$$M = \frac{\log.(1+n)}{N \log.(1+n)}. \qquad (D)$$

$M = \dfrac{1}{(a-1) - \frac{1}{2}(a-1)^2 + \frac{1}{3}(a-1)^3 - \&c.}$, is called the *modulus* of the system of logarithms whose base is a.

From (12) we see that, *the logarithms of any particular system is equal to the Napierian logarithm multiplied by the modulus of that porticular system*

(224.) *A new demonstration of the* LOGARITHMIC THEOREM, *and of the* BINOMIAL THEOREM, *for negative and fractional exponents.*

BY PROF. MARCUS CATLIN, A. M., A. S. S., HAMILTON COLLEGE, N. Y.

Let $1+y=$ any number. Assume

$$(1+y) = \text{P}a + by + cy^2 + dy^3 + \&c. \qquad (1)$$

where

$a + by + cy^2 + dy^3 + \&c. =$ the log. $(1+y)$; a, b, c, &c., being independent of n in the following binomial

$$(1+y)^n = \text{A} + \text{B}y + \text{C}y^2 + \text{D}y^3 + \&c. \qquad (2)$$

Equations (1) and (2) give

$$(1+y)^n = \text{P}na + nby + ncy^2 + ndy^3 + \&c. \qquad (3)$$

or,

$$\text{A} + \text{B}y + \text{C}y^2 + \text{D}y^3 + \&c. = \text{P}na + nby + ncy^2 + ndy^3 + \&c. \qquad (4)$$

or, $\begin{aligned} & 1 + (\text{A}-1+\text{B}y+\text{C}y^2+\text{D}y^3+\&c.) \\ & = \text{P}na + nby + ncy^2 + ndy^3 + \&c. \end{aligned} \qquad (5)$

LOGARITHMS.

But, by equation (1), the last expression may be written in this form

$$1+(A-1+By+Cy^2+Dy^3+\&c.) = \\ Pa+b(A-1+By+Cy^2+)+c(A-1+By+Cy^2+)^2+\&c. \quad (6)$$

By equating the second members of (5) and (6), we shall have

$$\begin{aligned} na+nby+ncy^2+ndy^3+\&c. &= \\ a+b(A-1+By+Cy^2+Dy^3+\&c.) \\ +c(A-1+By+Cy^2+Dy^3+\&c.)^2+\&c. \end{aligned} \quad (7)$$

Since a is independent of y, if we suppose $y=0$, equation (1) gives $a=0$; in a similar manner from (2) we find $A=1$; hence, equation (7) becomes

$$\begin{aligned} nby+ncy^2+ndy^3+\&c. &= \\ b(By+Cy^2+Dy^3+\&c.)+c(By+Cy^2+Dy^3+\&c.)^2 \\ +d(By+Cy^2+Dy^3+\&c.)^3+\&c. \end{aligned} \quad (8)$$

Equating the coefficients of the several powers of y with zero, we find

$$Bb-nb=0, \quad \therefore B=n, \quad (9)$$

$$B^2c-nc+cb=0, \quad \therefore -c=\frac{B^2c-nc}{b}, \quad (10)$$

$$nd-bD-2BCc-B^3d=0, \quad \therefore D=\frac{nd-2BCc-B^3d}{b}, \quad (11)$$

&c., &c.

Now, since a, b, c, &c., are independent of n, by assuming n successively equal to 1, 2, 3, &c., and substituting the corresponding values of C, D, &c., derived from (2), the equations (10), (11), &c. will give, after the necessary reductions,

$$b=1, \ c=-\tfrac{1}{2}, \ d=\tfrac{1}{3}, \ \&c.$$

Hence, by (1),

$$\log.(1+y)=(y-\tfrac{1}{2}y^2+\tfrac{1}{3}y^3-\tfrac{1}{4}y^4+\&c.) \quad (12)$$

P, being the base of the system. If we put $P = P'^m$, equation (12) will become

$$\log. (1+y) = m(y - \tfrac{1}{2}y^2 + \tfrac{1}{3}y^3 - \tfrac{1}{4}y^4 + \&c.), \qquad (13)$$

where P′ is the base, and m the modulus. Hence, the logarithmic theorem is proved.

Again, substituting the values of b, c, d, &c., in equations (10), (11), &c., they become

$$c = \frac{n(n-1)}{1.2}, \qquad (14)$$

$$D = \frac{n(n-1)(n-2)}{1.2.3}, \qquad (15)$$

&c., &c.

Substituting the values of A, B, C, D, &c., in (2), it becomes

$$(1+y)^n = 1 + ny + \frac{n(n-1)}{1.2} \cdot y^2 + \frac{n(n-1)(n-2)}{1.2.3} \cdot y^3 + \&c. \quad (16)$$

Hence, the binomial theorem is proved for fractional and negative exponents. The theorem for positive integral exponents is, of course, assumed in this demonstration.[*]

(225.) We will now proceed to calculate some Napierian

---

[*] This is taken from the second number of "*The Mathematical Miscellany*," a work which was devoted to pure mathematics, published at College Point, N. Y., and edited by PROF. C. GILL: after reaching its eighth number it was, in 1839, discontinued. We are happy to say that it has again made its appearance under the name of "*The Cambridge Miscellany*," edited by PROF. B. PIERCE, the first number of which was published in April, 1842: its columns are now open to questions in physico-mathematics, as well as to those of pure mathematics.

logarithms. If we take n negative in the formula (C'), we shall have

$$N\log.(1-n) = -n - \tfrac{1}{2}n^2 - \tfrac{1}{3}n^3 - \tfrac{1}{4}n^4 - \&c. \qquad (1)$$

Subtracting (1) from (C'), we have

$$\begin{aligned}N\log.(1+n) - N\log.(1-n) &= N\log.\frac{1+n}{1-n} \\ &= 2(n + \tfrac{1}{3}n^3 + \tfrac{1}{5}n^5 + \tfrac{1}{7}n^7 + \tfrac{1}{9}n^9 + \&c.)\end{aligned} \qquad (2)$$

If we assume $n = \frac{1}{2p+1}$, we shall find $\frac{1+n}{1-n} = \frac{p+1}{p}$, and (2) will become

$$N\log.\frac{p+1}{p} = 2\left\{\frac{1}{2p+1} + \frac{1}{3(2p+1)^3} + \frac{1}{5(2p+1)^5} + \&c.\right\} \qquad (3)$$

or, which is the same thing,

$$N\log.(p+1) = N\log.p + 2\left\{\frac{1}{2p+1} + \frac{1}{3(2p+1)^3} + \frac{1}{5(2p+1)^5} + \&c.\right\} \qquad (E)$$

If we take $p=1$, formula (E) becomes

$$N\log. 2 = 2\left\{\frac{1}{3} + \frac{1}{3.3^3} + \frac{1}{5.3^5} + \frac{1}{7.3^7} + \&c.\right\}$$

```
        3|2
3²=9|0.66666666÷ 1=0.66666666
    9|0.07407407÷ 3=0.02469136
    9|0.00823045÷ 5=0.00164609
    9|0.00091449÷ 7=0.00013064
    9|0.00010161÷ 9=0.00001129
    9|0.00001129÷11=0.00000103
    9|0.00000125÷13=0.00000010
    9|0.00000014÷15=0.00000001
     |0.00000001
                  ─────────────
                  0.69314718=N log. 2.
```

LOGARITHMS.

Take $p=4$ in formula (E), and we get

$$N\log.5 = N\log.4 + 2\left\{\frac{1}{9} + \frac{1}{3.9^3} + \frac{1}{5.9^5} + \frac{1}{7.9^7} + \&c.\right\}$$

But $N\log.5 = N\log.10 - N\log.2$;
also, $N\log.4 = 2\,N\log.2$.

Hence, substituting these values of $N\log.5$ and $N\log.4$, in the above expression and we get, after transposing,

$$N\log.10 = 3\,N\log.2 + 2\left\{\frac{1}{9} + \frac{1}{3.9^3} + \frac{1}{5.9^5} + \frac{1}{7.9^7} + \&c.\right\}$$

Executing the calculation, for the sum of the series, as in the above example, omitting the cyphers on the left, we obtain the following:

```
9|2
9|0.22222222 ÷ 1 = 0.22222222
9|   2469136
9|    274348 ÷ 3 =      91449
9|     30483
9|      3387 ÷ 5 =        677
9|       376
9|        42 ÷ 7 =          6
0|         5                 ─────────
                   0.22314354 = sum of series.
         3 N log. 2 = 2.07944154
                    ─────────
                    2.30258508 = N log. 10.
```

We are now prepared to find the modulus of the Briggean system. Since the base of the Briggean system is 10, and the logarithm of the base of any system is 1, we have log. $10 = 1$; formula (D) shows, *that the common logarithm of any number divided by the Napierian logarithm is equal to the modulus of the common system.*

Hence,

$$M = \frac{\log. 10}{N \log. 10} = \frac{1}{2.30258508} = 0.43429448.$$

This value, when carried to 35 decimal places, is

$$M = 0.43429448190325182765112891891660508.$$

We will now proceed to calculate common logarithms.

Since all numbers are composed of a certain number of *prime* factors, and since the logarithm of any number is equal to the sum of the logarithms of all its factors, it follows that it will be necessary only to calculate the logarithms of prime numbers.

By equation (12), Art. 223, we see that the Napierian logarithm multiplied by M, gives the common logarithm.

Hence,

$$\log. 2 = N \log. 2 \times M =$$
$$0.69314718 \times 0.43429448 = 0.30103000.$$

The logarithm of 10 we know to be 1, therefore the

$$\log. 5 = \log. \frac{10}{2} = 1 - \log. 2 = 0.69897000.$$

Formula (E), when adapted to common logarithms, becomes

$$\log. (p+1) =$$
$$\log. p + 2M \left\{ \frac{1}{2p+1} + \frac{1}{3(2p+1)^3} + \frac{1}{5(2p+1)^5} + \&c. \right\}$$

or

$$\log. (p+1) = \log. p +$$
$$0.86858896 \left\{ \frac{1}{2p+1} + \frac{1}{3(2p+1)^3} + \frac{1}{5(2p+1)^5} + \&c. \right\} \quad (F)$$

Take $p=2$ in (F), and we get

$$\log. 3 = \log. 2 + 0.86858896 \left\{ \frac{1}{5} + \frac{1}{3.5^3} + \frac{1}{5.5^5} + \frac{1}{7.5^7} + \&c. \right\}$$

LOGARITHMS.

```
        5|0.86858886
 5²=25|0.17371779÷1=0.17371779
       25|     694871÷3=     231623
       25|      27795÷5=       5559
       25|       1112÷7=        159
       25|         44÷9=          5
        |          2
                              ───────────
                              0.17609125 =sum of series.
                   log. 2=0.30103000
                              ───────────
                              0.47712125=log. 3.
```

If, in (F), we make $p=49$, we get

log. 50 =

$\quad$ log. 49 + 0.86858895 $\left\{ \dfrac{1}{99} + \dfrac{1}{3.(99)^2} + \dfrac{1}{5.(99)^5} + \&c. \right\}$

and since log. 50 = log. 10 + log. 5, and log. 49 = 2 log. 7, we have by substitution and transposition,

2 log. 7 =

log. 10 + log. 5 − 0.86858896 $\left\{ \dfrac{1}{99} + \dfrac{1}{3.(99)^3} + \dfrac{1}{5.(99)^5} + \&c. \right\}$

Calculating the series we find

```
              99|0.86858896
(99)²=9801|    877362÷1=0.00877362
                    89÷3=         29
                              ───────────
                              0.00877391=sum of series.

            log. 5=0.69897000
            log. 10=1.
                              ───────────
                              1.69897000
                              0.00877391
                              ───────────
                              1.69019609=2 log. 7
                              0.84509804=log. 7.
```

We might have calculated the log. 7 by substituting 6 for p in (F), but the operation would have been more lengthy than the above.

The next prime in order is 11, to find its log., we make, in equation (F), $p=99$, observing that log. $100=2$, also, that

$$\log. 99 = \log. 9 + \log. 11 = 2 \log. 3 + \log. 11,$$

we thus obtain

$$2 = 2 \log. 3 + \log. 11 + 0.86858896 \left\{ \frac{1}{199} + \frac{1}{3(199)^3} + \&c. \right\}$$

or by transposing, it becomes

$$\log. 11 = 2 - 2 \log. 3 - 0.86858896 \left\{ \frac{1}{199} + \frac{1}{3(199)^3} + \&c. \right\}$$

```
199|0.86858896
 39601|    436477÷1=0.00436477
           11÷3=        4
                  ─────────────
                  0.00436481=sum of series.
         2 log. 3=0.95424250
                  ─────────────
                  0.95860731
2.00000000
0.95860731
─────────────
1.04139269=log. 11.
```

To find the log. of the next prime 13, we assume $p=1000$ in equation (F), and obtain

log. 1001=

$$\log. 1000 + 0.86858896 \left\{ \frac{1}{2001} + \frac{1}{3(2001)^3} + \&c. \right\}$$

Now, since

$1001 = 7 \times 11 \times 13$, log. $1001 = \log. 7 + \log. 11 + \log. 13.$

Hence,

log. 13 =

$3 - \log. 7 - \log. 11 + 0.86858896 \left\{ \dfrac{1}{2001} + \dfrac{1}{3(2001)^3} + \&c. \right\}$

2001) 0.86858896
　　　..43407 = sum of series.
　　　　　　log. 7 = 0.84509804
　　　　　　log. 11 = 1.04139269

3.00043407　　　　　　1.88649073
1.88649073

1.11394334 = log. 13.

We might proceed in this way until we should have calculated the logarithms of all the prime members within the limits of the tables.

(226.) We have already (Art. 220) said that the base a of the Napierian system of logarithms satisfies the following equation:

$$(a-1) - \tfrac{1}{2}(a-1)^2 + \tfrac{1}{3}(a-1)^3 - \tfrac{1}{4}(a-1)^4 + \&c. = 1. \quad (1)$$

From example 3, page 231, we see that if we have

$$(y-1) - \dfrac{(y-1)^2}{2} + \dfrac{(y-1)^3}{3} + \dfrac{(y-1)^4}{4} + \&c. = x, \quad (2)$$

then will

$$y = 1 + x + \dfrac{x^2}{2} + \dfrac{x^3}{2.3} + \dfrac{x^4}{2.3.4} + \&c. \quad (3)$$

Equation (2) will agree with (1), when $y = a$; and $x = 1$. Making these changes in (3), we find

$$a = 1 + 1 + \dfrac{1}{2} + \dfrac{1}{2.3} + \dfrac{1}{2.3.4} + \dfrac{1}{2.3.4.5} + \&c. \quad (4)$$

This series may be summed as follows:

LOGARITHMS.

```
 1 | 1
 2 | 1
 3 | 0.5
 4 | 0.16666666
 5 |   4166666
 6 |    833333
 7 |    138888
 8 |     19841
 9 |      2480
10 |       276
11 |        28
   |         2
```

2.71828180 = base of Napierian logarithms.

This value when extended to 35 decimals is found to be

$e = 2.71828182845904523536028747135266249$.

EXPONENTIAL THEOREM.

(227.) The above theorem makes known the law of the development of a^x according to the ascending powers of x.

To determine this law, we will assume

$$a^x = 1 + Ax + Bx^2 + Cx^3 + Dx^4 + \&c., \quad (1)$$

both members of which become 1 when $x = 0$.

Changing x into y, in (1), and we have

$$a^y = 1 + Ay + By^2 + Cy^3 + Dy^4 + \&c. \quad (2)$$

Subtracting (2) from (1), and actually dividing the right-hand member by $x - y$, we obtain

$$\frac{a^x - a^y}{x - y} = A + B(x+y) + C(x^2 + xy + y^2) \\ + D(x^3 + x^2 y + xy^2 + y^3) + \&c. \quad (3)$$

Writing $x - y$ for x in (1), and it becomes

$$a^{x-y} = 1 + A(x-y) + B(x-y)^2 + C(x-y)^3 + \&c. \quad (4)$$

LOGARITHMS.

Transposing the 1, and multiplying by a^y, we get
$$a^y(a^{x-y}-1)=a^y\{A(x-y)+B(x-y)^2+C(x-y)^3+\&c.\} \quad (5)$$

Divided (5) by $x-y$, after replacing its left-hand member by its equivalent value a^x-a^y, we find

$$\frac{a^x-a^y}{x-y}=a^y\{A+B(x-y)+C(x-y)^2+D(x-y)^3+\&c.\} \quad (6)$$

Equating the right-hand members of (3) and (6), we have

$$\left\{\begin{array}{c} A+B(x+y)+C(x^2+xy+y^2) \\ +D(x^3+x^2y+xy^2+y^3)+\&c. \\ =a^y\{A+B(x-y)+C(x-y)^2+D(x-y)^3+\&c.\} \end{array}\right\} \quad (7)$$

This is true for all values of x and y.

When $y=x$, it becomes

$$A+2Bx+3Cx^2+4Dx^3+\&c.=a^x\cdot A. \quad (8)$$

For a^x, substituting its value equation (1), we find

$$\left.\begin{array}{c} A+2Bx+3Cx^2+4Dx^3+\&c. \\ =A+A^2x+ABx^2+ACx^3+\&c. \end{array}\right\} \quad (9)$$

Equating the coefficients of the like powers of x, Art. 183, we find

$$A=A,$$
$$2B=A^2,$$
$$3C=AB,$$
$$4D=AC,$$
$$\&c., \quad \&c.$$

From which we find

$$A=A,$$
$$B=\frac{A^2}{2},$$
$$C=\frac{A^3}{3},$$
$$D=\frac{A^4}{4},$$
$$\&c., \quad \&c.$$

Hence, (1) becomes

$$a^x = 1 + Ax + \frac{A^2 x^2}{2} + \frac{A^3 x^3}{2.3} + \frac{A^4 x^4}{2.3.4} + \&c. \qquad (10)$$

It now remains to find the value of A.

For this purpose put $1+b=a$, and we have $a^x=(1+b)^x$, which by the Binomial Theorem becomes

$$(1+b)^x = 1 + \frac{xb}{1} + \frac{x(x-1)b^2}{1.2} + \frac{x(x-1)(x-2)b^3}{1.2.3} + \&c. \qquad (11)$$

Performing the multiplications indicated, we find the coefficient of the first power of x to be

$$\frac{b}{1} - \frac{b^2}{2} + \frac{b^3}{3} - \frac{b^4}{4} + \&c.,$$

or resubstituting $a-1$ for b, it becomes

$$(a-1) - \frac{(a-1)^2}{2} + \frac{(a-1)^3}{3} - \frac{(a-1)^4}{4} + \&c.$$

Therefore

$$A = (a-1) - \frac{(a-1)^2}{2} + \frac{(a-1)^3}{3} - \frac{(a-1)^4}{4} + \&c. \qquad (12)$$

If in formula (C'), we put $a-1$ for n, we shall find

$$N \log. a = (a-1) - \frac{(a-1)^2}{2} + \frac{(a-1)^3}{3} - \frac{(a-1)^4}{4} + \&c.$$

Hence,

$$A = N \log. a. \qquad (13)$$

This value of A substituted in (10), gives

$$a^x = 1 + N \log. a . x + \frac{(N \log. a)^2 . x^2}{2} + \frac{(N \log. a)^3 . x^3}{2.3} + \quad (A)$$

When $a = e = 2.7182818\&c.$,
then $N \log. a = N \log. e = 1$,
and (A) becomes

$$e^x = 1 + x + \frac{x^2}{2} + \frac{x^3}{2.3} + \frac{x^4}{2.3.4} + \&c. \qquad (B)$$

LOGARITHMS.

APPLICATION OF LOGARITHMS.

(228.) By the aid of a table of logarithms, we can easily perform the following operations:

1. Find the value of $\dfrac{3.75 \times 1.06}{365}$ by logarithms.

Recollecting (Art. 216) that the logarithm of the product of several factors is equal to the sum of their respective logarithms; and (Art. 218) the logarithm of the quotient of one quantity divided by another is equal to the logarithm of the dividend diminished by the logarithm of the divisor, we find for the logarithm of our expression

$$\log. \frac{3.75 \times 1.06}{365} = \log. 3.75 + \log. 1.06 - \log. 365.$$

By the tables we have

$$\log. 3.75 = 0.5740313$$
$$\log. 1.06 = 0.0253059$$
$$\overline{0.5993372}$$
$$\log. 365 = 2.5622929$$
$$\overline{\log. 0.01089 = \overline{2}.0370443}$$

Therefore, the above expression is nearly equal to 0.01089.

2. Find the 11th root of 11, that is, the value of $\sqrt[11]{11}$.

Taking the logarithm of this expression, we find

$\log. \sqrt[11]{11} = \tfrac{1}{11}$ of $\log. 11 = \tfrac{1}{11}$ of $1.0413927 = 0.0946721$
$ = \log. 1.24357$ &c.

Therefore, $\sqrt[11]{11} = 1.24357.$

3. What is the value of $\dfrac{8^5 \times \sqrt[3]{7}}{\sqrt[5]{6}}$?

$5 \times \log. 8 + \tfrac{1}{3} \log. 7 - \tfrac{1}{5} \log. 6 = 4.51545 + 0.2816993$
$ - 0.1556302 = 4.6415191 = \log. 43794.53.$

Therefore, our expression is equivalent to 43804.53.

LOGARITHMS.

(229.) The above examples will show the great advantage of logarithms in abridging arithmetical labor. In the higher parts of analysis, the use of logarithms is indispensable. It would not be difficult to propose questions, which by logarithms, might be wrought in a few moments, but if wrought by arithmetical rules, would require years. The following example will illustrate the above remark.

How many figures will be required to express 9^{9^9} ?

The exponent of the above expression is

$9^9 = 387420489 \quad \therefore \quad 9^{9^9} = 9^{387420489}$.

Putting it into logarithms, we have

$\log. 9^{387420489} = 387420489 \times \log. 9 =$
$387420489 \times 0.954242509439 = 369693099.634$ &c.

Hence, the number answering to this logarithm must consist of 369693100 figures. This number, if printed, would fill upwards of 256 volumes of 400 pages each, allowing 60 lines to a page, and 60 figures to a line.

EXPONENTIAL EQUATIONS.

(230.) An exponential equation is one where the unknown quantity enters as an exponent.

Thus, $\quad a^x = b ; \quad x^x = c ;$ &c.,

are exponential equations.

(231.) When the equation is of the form $a^x = b$, we find, by taking the logarithm of both members, $x \cdot \log. a = \log. b$.

Therefore, $\quad x = \dfrac{\log. b}{\log. a}$.

(232.) When the exponential is of this form $x^x = c$, we must find the value of x by the following double position.

LOGARITHMS.

RULE.

Find by trial two numbers as near the value of x as possible, and substitute them successively for x, then, as the difference of the results is to the difference of the two assumed numbers, so is the difference of the true result, and either of the former, to the difference of the true number and the supposed one belonging to the result last used; this difference, therefore, being added to the supposed number, or subtracted from it, according as it is too little or too great, will give the true value nearly.

And if this near value be substituted for x, as also the nearest of the first assumed numbers, unless a number still nearer be found, and the above operations be repeated, we shall obtain a still nearer value of x; and in this way we may continually approximate to the true value of x.

EXAMPLES.

1. Given $x^x=100$, to find an approximate value of x.

The above equation when put into logarithms, becomes

$$x \times \log. x = \log. 100 = 2. \qquad (1)$$

By a few trials we find the value of x to fall between 3 and 4. If we substitute, in succession, 3 and 4 in (1), we shall find

$$3 \times \log. 3 = 1.4313639$$
$$4 \times \log. 4 = 2.4082400$$

$$0.9768761 = \text{diff. of results.}$$

$$0.9768761 : 1 :: 0.4082400 : 0.418.$$

Hence, $\quad 4 - 0.418 = 3.582 = x$ nearly.

Upon trial, this value is found to be rather too small, whilst 3.6 is rather too great; therefore, substituting each of these in succession in (1), we find

$3.582 \times \log. 3.582 = 1.9848779$
$3.6 \times \log. 3.6 = 2.0026890$

$\overline{0.0178111} = $ diff. of result.

$0.0178111 : 0.018 :: 0.0026890 : 0.002717$.

Hence,
$3.6 - 0.002717 = 3.597283 = x$ nearly.

2. Given $x^x = 5$, to find an approximate value of x.

Ans. $x = 2.1289$.

3. Given $x^x = 2000$, to find an approximate value of x.

Ans. $x = 4.8278$.

COMPOUND INTEREST AND ANNUITIES BY LOGARITHMS.

(233.) *Interest* is money paid by the borrower for the use of the money borrowed.

It is estimated at a certain *rate per cent. per annum*, that is, a certain number of dollars for the use of $100 for one year.

The sum upon which the interest is computed, is called the *principal*.

The principal when increased by the interest, is called the *amount*.

When the interest of a given principal is paid at the end of each year, it is called *simple* interest; but when the interest due, at the end of each year, goes to increase the principal, it is called *compound* interest.

The *present worth*, at compound interest, of a given debt due at some future time, is such a sum as being put out at compound interest, will in the given time amount to the debt.

An annuity is a fixed sum which is paid periodically, for a certain length of time.

(234.) In our calculations we shall use the following notation:

$p=$ the principal.
$r=$ the interest of \$1 for one year.
$R=\$1+r=$ the amount of \$1 for one year.
$a=$ the amount of the given principal.
$A=$ an annuity.
$a'=$ the amount of a given annuity.
$P=$ the present worth of a given annuity.
$n=$ the time in years.

Since $\$1+r=R$ is the amount of \$1 for one year, it follows, that the amount of a given principal, p will in the same time be pR, and this being considered as a new principal, will in the next year amount to $pR \times R = pR^2$, which in turn will the next year amount to $pR^2 \times R = pR^3$, and so on.

Hence,

$pR =$ amount for 1 year.
$pR^2=$ amount for 2 years.
$pR^3=$ amount for 3 years.
$pR^4=$ amount for 4 years.
- - - - - - - -
- - - - - - - -
$pR^n=$ amount for n years.

Therefore, we have this relation

$$a=pR^n,$$

which, in logarithms, becomes

$$\log. a = \log. p + n \log. R. \qquad (1)$$

(235.) When an annuity is left unpaid, for n years, it is obvious that the annuity due at the end of the first year, must be on interest $n-1$ years, and must therefore amount to AR^{n-1}, the annuity due at the end of the second year will be on interest $n-2$ years and will therefore amount to AR^{n-2}, and so on, hence, the amount of the annuity at the end of n years will be

$$A(R^{n-1}+R^{n-2}+ \ - \ - \ - \ - \ R+1).$$

The geometrical progression within the parenthesis being summed, we have, after substituting r for $R-1$,

$$a'=A\left(\frac{R^n-1}{r}\right). \qquad (2)$$

We have said that the present worth of a debt is such a sum as being put out at interest, will in the given time amount to the debt, hence we have

$$PR^n=A\left(\frac{R^n-1}{r}\right) \qquad (2')$$

from which we find

$$P=\frac{A\left(1-\frac{1}{R^n}\right)}{r}=\frac{a'}{R^n}. \qquad (3)$$

When the annuity is continued *forever*, the value of n becomes infinite, making this substitution in (3), we find

$$P=\frac{A}{r}. \qquad (4)$$

EXAMPLES.

1. How much will $875 amount to in 12 years, at 6 per cent., compound interest?

In this example, we have

$$p=875; \ n=12; \ R=1.06;$$

and we are required to find a.

Substituting these values in equation (1), we have

$$\log. a=\log. 875+12 \log. 1.06.$$

By actually consulting a table of logarithms, we find

LOGARITHMS.

$$\log. 875 = 2.9420081$$
$$12 \log. 1.06 = 0.3036708$$

$$\log. a = 3.2456789.$$

Therefore, $\qquad a = \$1760.672.$

2. What principal will, in 10 years at 5 per cent., amount to $1000?

By transposition, equation (1) becomes
$$\log. p = \log. a - n \log. R.$$
Substituting for a, n and R their given values, we have
$$\log. p = \log. 1000 - 10 \log. 1.05,$$
$$\therefore \log. p = 3 - 0.2118930 = 2.7881070.$$
And, $\qquad p = \$613.913.$

3. At what rate per cent. will $100 in 16 years amount to $160?

Equation (1) gives
$$\log. R = \frac{\log. a - \log. p}{n},$$
which, in this example, becomes
$$\log. R = \frac{2.2041200 - 2}{16} = 0.0127575,$$
$$\therefore R = 1.02981.$$
Therefore, the per cent. is 2.981, or nearly 3 per cent.

4. In how many years will $460 at 7 per cent. amount to $1000?

Again, equation (1) gives
$$n = \frac{\log. a - \log. p}{\log. R},$$
which, in this example, becomes
$$n = \frac{3 - 2.6627578}{0.0293838} = 11.477 \text{ years nearly.}$$

LOGARITHMS.

5. What is the amount of an annuity of $200, which has remained unpaid 14 years, at 6 per cent., compound interest?

Equation (2), when put into logarithms, becomes

$$\log. a' = \log. A + \log. (R^n - 1) - \log. r.$$

In the present example

$$r = 0.06 \; ; \; R = 1.06 \; ; \; A = 200 \; ; \; n = 14.$$
$$\log. R^n = n \log. R = 0.3542826,$$
$$\therefore R^n = 2.2609 \text{ and } R^n - 1 = 1.2609.$$

Hence,

$$\log. a' = 2.3010300 + 0.1006806 - \overline{2}.7781513,$$

and $\log. a' = 3.6235593.$

Therefore, $a' = \$4203$ nearly.

6. What is the present worth of the above annuity?

Equation (3) gives

$$\log. P = \log. a' - n \log. R.$$

In this particular case, we have

$$\log. P = 3.6235593 - 0.3542826 = 3.2692767.$$

And $P = \$1858.988.$

7. What is the present worth of an annuity of $100, to continue forever, at 7 per cent.?

By equation (4), which is $P = \dfrac{A}{r}$, we find

$$P = \frac{\$100}{0.07} = \$1428.571.$$

8. A debt, due at the present time, amounting to $1200, is to be discharged in seven yearly and equal payments. What is the amount of one of these payments, if the interest is calculated at 4 per cent.?

In this example, we have given the present worth of an

annuity, the time of its continuance and the rate of interest, to find the annuity.

Equation (2'), by a slight reduction, becomes
$$A = \frac{PrR^n}{R^n - 1},$$
which, in logarithms, is

log. A = log. P + log. r + n log. R − log. (R^n − 1).

If we take
$$P = \$1200 \; ; \; r = 0.04 \; ; \; R = 1.04 \; ; \; n = 7,$$
we shall find
$$A = \$199.931.$$

9. In what time will a given principal, at compound interest, amount to m times the principal?

Under example 4, we have the formula
$$n = \frac{\log. a - \log. p}{\log. R}.$$

To make this agree with the present case, we must, for a, write mp, by which means it becomes
$$n = \frac{\log. m}{\log. R}.$$

(236.) When the interest, instead of being added to the principal at the end of each year, is added at any other regular period, as half yearly, quarterly, &c., n must be considered as standing for the number of those periods, and r will be the interest for one of those periods.

10. What is the amount of $100, for 3 years, at 6 per cent. per annum, when the interest is added at the end of every 6 months?

Equation (1), when adapted to the present example, becomes

$$\log. a = \log. 100 + 6 \log. 1.03,$$
from which we find
$$a = \$119.405.$$

11. If the interest of \$1, for the xth part of a year, is $\dfrac{r}{x}$, what will be the amount of \$1 for n years, when $x = \infty$?

The formula for the amount will, in this case, be
$$a = \left(1 + \frac{r}{x}\right)^{nx}.$$

Expanding the right-hand member by the Binomial Theorem, we find
$$a = 1 + nx \cdot \frac{r}{x} + \frac{nx(nx-1)}{1.2} \cdot \frac{r^2}{x^2} + \frac{nx(nx-1)(nx-2)}{1.2.3} \cdot \frac{r^3}{x^3} + \&c.$$

When $x = \infty$, this becomes
$$a = 1 + nr + \frac{n^2 r^2}{1.2} + \frac{n^3 r^3}{1.2.3} + \frac{n^4 r^4}{1.2.3.4} + \&c.$$

Comparing this with formula (B), Art. 227, we have
$$a = e^{nr}.$$

Using the common logarithms, we have
$$\log. a = nr \times 0.4342944819.$$

(237.) Before closing this chapter, we will show how formulas 17, 18, 19, and 20 of Geometrical Progression were found.

By taking the logarithm of both members of No. 1, as given in the table under Art. 178, we have
$$\log. l = \log. a + (n-1) \log. r.$$

This gives
$$n - 1 = \frac{\log. l - \log. a}{\log. r},$$

LOGARITHMS.

or
$$n = \frac{\log. l - \log. a}{\log. r} + 1,$$

which agrees with No. 17.

No. 5 is readily put in the following form:
$$a + (r-1)s = ar^n.$$

Taking the logarithm, we have
$$\log. [a + (r-1)s] = \log. a + n \log. r,$$

from which we readily get
$$n = \frac{\log. [a + (r-1)s] - \log. a}{\log. r},$$

which agrees with No. 18.

No. 12 may take the following form:
$$a(s-a)^{n-1} = l(s-l)^{n-1},$$

which in logarithms is
$$\log. a + (n-1) \log. (s-a) = \log. l + (n-1) \log. (s-l),$$

which gives
$$n = \frac{\log. l - \log. a}{\log. (s-a) - \log. (s-l)} + 1,$$

which agrees with No. 19.

Again, No 16 may be written as follows:
$$r(s-l)r^{n-1} - sr^{n-1} + l = 0,$$

which is readily reduced to
$$[rl - (r-1)s]r^{n-1} = l.$$

Taking the logarithms, we have
$$\log. [rl - (r-1)s] + (n-1) \log. r = \log. l.$$

From this we find
$$n = \frac{\log. l - \log. [rl - (r-1)s]}{\log. r} + 1,$$

which is the same as No. 20.

LOGARITHMS

I HAVE inserted this chapter on logarithms because I consider a knowledge of them very essential to the education of any engineer.

Definition. — A logarithm is the power to which we must raise a given base to produce a given number. Thus, suppose we choose 10 as our base, we will say that 2 is the logarithm of 100, because we must raise 10 to the second power — in other words, square it — in order to produce 100. Likewise 3 is the logarithm of 1000, for we have to raise 10 to the third power (thus, 10^3) to produce 1000. The logarithm of 10,000 would then be 4, and the logarithm of 100,000 would be 5, and so on.

The *base* of the universally used *Common System* of logarithms is 10; of the *Naperian* or *Natural System*, e or 2.7. The latter is seldom used.

We see that the logarithms of such numbers as 100, 1000, 10,000, etc., are easily detected; but suppose we have a number such as 300, then the difficulty of finding its logarithm is apparent. We have seen that 10^2 is 100, and 10^3 equals 1000, therefore the number 300, which lies between 100 and 1000, must have a logarithm which lies between the logarithms of 100 and 1000, namely 2 and 3 respectively. Reference to a table of logarithms at the end of this book, which we will explain later, shows that the logarithm of 300 is 2.4771, which means that 10 raised to the 2.4771ths power will give 300. The *whole number* in a logarithm, for example the 2 in the above case, is called the *characteristic;* the decimal part of the logarithm, namely, .4771, is called the *mantissa*. It will be hard for the student to understand at first what is meant by raising 10 to a fractional part of a power, but he should not worry about this at the present time; as he studies more deeply into mathematics the notion will dawn on him more clearly.

We now see that every number has a logarithm, no matter how large or how small it may be; every number can be produced by raising 10 to some power, and this power is what we call the *logarithm* of the number. Mathematicians have carefully worked out and tabulated the logarithm of every number, and by reference to these tables we can find the logarithm corresponding to any number, or vice versa. A short table of logarithms is shown at the end of this book.

Now take the number 351.1400; we find its logarithm is 2.545,479. Like all numbers which lie between 100 and 1000 its characteristic is 2. The numbers which lie between 1000 and 10,000 have 3 as a characteristic; between 10 and 100, 1 as a characteristic. We there-

LOGARITHMS

fore have the rule that *the characteristic is always one less than the number of places to the left of the decimal point.* Thus, if we have the number 31875.12, we immediately see that the characteristic of its logarithm will be 4, because there are five places to the left of the decimal point. Since it is so easy to detect the characteristic, it is never put in logarithmic tables, the *mantissa* or *decimal* part being the only part that the tables need include.

If one looked in a table for a logarithm of 125.60, he would only find .09,899. This is only the *mantissa* of the logarithm, and he would himself have to insert the characteristic, which, being one less than the number of places to the left of the decimal point, would in this case be 2 ; therefore the logarithm of 125.6 is 2.09,899.

Furthermore, the *mantissæ* of the logarithms of 3.4546, 34.546, 345.46, 3454.6, etc., are all exactly the same. The characteristic of the logarithm is the only thing which the decimal point changes, thus:

$$\log 3.4546 = 0.538{,}398{,}$$
$$\log 34.546 = 1.538{,}398{,}$$
$$\log 345.46 = 2.538{,}398{,}$$
$$\log 3454.6 = 3.538{,}398{,}$$
$$\text{etc.}$$

Therefore, in looking for the logarithm of a number, first put down the *characteristic* on the basis of the above rules, then look for the *mantissa* in a table, neglecting the position of the decimal point altogether. Thus, if we are looking for the logarithm of .9840, we first write down the characteristic, which in this case would be -1 (there are no places to the left of the decimal point in this case, therefore one less than none is -1). Now look in a table of logarithms for the mantissa which corresponds to .9840, and we find this to be .993,083; therefore

$$\log .9840 = -1.993,083.$$

If the number had been 98.40 the logarithm would have been $+1.993,083$.

When we have such a number as .084, the characteristic of its logarithm would be -2, there being one less than no places at all to the left of its decimal point; for, even if the decimal point were moved to the right one place, you would still have no places to the left of the decimal point; therefore

$$\log .00,386 = -3.586,587,$$
$$\log 38.6 = 1.586,587,$$
$$\log 386 = 2.586,587,$$
$$\log 386,000 = 5.586,587.$$

Interpolation. — Suppose we are asked to find the logarithm of 2468; immediately write down 3 as the characteristic. Now, on reference to the logarithmic

LOGARITHMS

table at the end of this book, we see that the logarithms of 2460 and 2470 are given, but not 2468. Thus:

$$\log 2460 = 3.3909,$$
$$\log 2468 = ?$$
$$\log 2470 = 3.3927.$$

We find that the total difference between the two given logarithms, namely 3909 and 3927, is 18, the total difference between the numbers corresponding to these logarithms is 10, the difference between 2460 and 2468 is 8; therefore the logarithm to be found lies $\frac{8}{10}$ of the distance across the bridge between the two given logarithms 3909 and 3927. The whole distance across is 16. $\frac{8}{10}$ of 16 is 12.8. Adding this to 3909 we have 3921.8; therefore

$$\log \text{ of } 2468 = 3.39,218.$$

Reference to column 8 in the interpolation columns to the right of the table would have given this value at once.

Many elaborate tables of logarithms may be purchased at small cost which make interpolation almost unnecessary for practical purposes.

Now let us work backwards and find the number if we know its logarithm. Suppose we have given the logarithm 3.6201. Referring to our table, we see that the mantissa .6201 corresponds to the number 417; the characteristic 3 tells us that there must be four places to the left of the decimal point; therefore

3.6201 is the log of 4170.0.

Now, for interpolation we have the same principles aforesaid. Let us find the number whose log is -3.7304. In the table we find that

>log 7300 corresponds to the number 5370,
>log 7304 corresponds to the number ?
>log 7308 corresponds to the number 5380.

Therefore it is evident that

>7304 corresponds to 5375.

Now the characteristic of our logarithm is -3; from this we know that there must be two zeros to the right of the decimal point; therefore

>-3.7304 is the log of the number .005375.

Likewise

>-2.7304 is the log of the number .05375,
>.7304 is the log of the number 5.375,
>4.7304 is the log of the number 53,750.

Use of the Logarithm. — Having thoroughly understood the nature and meaning of a logarithm, let us investigate its use mathematically. It changes *multiplication* and *division* into *addition* and *subtraction; involution* and *evolution* into *multiplication* and *division.*

We have seen in algebra that

$$a^2 \times a^5 = a^{5+2}, \text{ or } a^7,$$

and that
$$\frac{a^8}{a^3} = a^{8-3}, \text{ or } a^5.$$

In other words, multiplication or division of like symbols was accomplished by adding or subtracting their exponents, as the case may be. Again, we have seen that

$$(a^2)^2 = a^4,$$
or $$\sqrt[3]{a^6} = a^2.$$

In the first case a^2 squared gives a^4, and in the second case the cube root of a^6 is a^2; to raise a number to a power you multiply its exponent by that power; to find any root of it you divide its exponent by the exponent of the root. Now, then, suppose we multiply 336 by 5380; we find that

$$\log \text{ of } 336 = 10^{2.5263},$$
$$\log \text{ of } 5380 = 10^{3.7308}.$$

Then 336×5380 is the same thing as $10^{2.5263} \times 10^{3.7308}$.

But $10^{2.5263} \times 10^{3.7308} = 10^{2.5263 + 3.7308} = 10^{6.2571}$.

We have simply added the exponents, remembering that these exponents are nothing but the logarithms of 336 and 5380 respectively.

Well, now, what number is $10^{6.2571}$ equal to? Looking in a table of logarithms we see that the mantissa .2571 corresponds to 1808; the characteristic 6 tells us that there must be seven places to the left of the decimal; therefore

$$10^{6.2571} = 1,808,000.$$

If the student notes carefully the foregoing he will see that in order to multiply 336 by 5380 we simply find their logarithms, add them together, getting another logarithm, and then find the number corresponding to this logarithm. Any numbers may be multiplied together in this simple manner; thus, if we multiply 217 × 4876 × 3.185 × .0438 × 890, we have

$$
\begin{aligned}
\log 217 &= 2.3365 \\
\log 4876 &= 3.6880 \\
\log 3.185 &= .5031 \\
\log .0438 &= -2.6415^* \\
\log 890 &= \underline{2.9494}
\end{aligned}
$$

Adding we get 8.1185

We must now find the number corresponding to the logarithm 8.1185. Our tables show us that

8.1185 is the log of 131,380,000.

Therefore 131,380,000 is the result of the above multiplication.

To divide one number by another we subtract the logarithm of the latter from the logarithm of the former; thus, 3865 ÷ 735:

$$
\begin{aligned}
\log 3865 &= 3.5872 \\
\log 735 &= \underline{2.8663} \\
& .7209
\end{aligned}
$$

The tables show that .7209 is the logarithm of 5.259; therefore

$$3865 \div 735 = 5.259.$$

* The −2 does not carry its negativity to the mantissa.

LOGARITHMS

As explained above, if we wish to square a number, we simply multiply its logarithm by 2 and then find what number the result is the logarithm of. If we had wished to raise it to the third, fourth or higher power, we would simply have multiplied by 3, 4 or higher power, as the case may be. Thus, suppose we wish to cube 9879; we have

$$\begin{array}{r} \log 9897 = 3.9947 \\ 3 \\ \hline 11.9841 \end{array}$$

11.9841 is the log of 964,000,000,000; therefore 9879 cubed = 964,000,000,000.

Likewise, if we wish to find the square root, the cube root, or fourth root or any root of a number, we simply divide its logarithm by 2, 3, 4 or whatever the root may be; thus, suppose we wish to find the square root of 36,850, we have

$$\log 36,850 = 4.5664.$$
$$4.5664 \div 2 = 2.2832.$$

2.2832 is the log. of 191.98; therefore the square root of 36,850 is 191.98.

The student should go over this chapter very carefully, so as to become thoroughly familiar with the principles involved.

MATHEMATICS

PROBLEMS

1. Find the logarithm of 3872.
2. Find the logarithm of 73.56.
3. Find the logarithm of .00988.
4. Find the logarithm of 41,267.
5. Find the number whose logarithm is 2.8236.
6. Find the number whose logarithm is 4.87175.
7. Find the number whose logarithm is −1.4385.
8. Find the number whose logarithm is −4.3821.
9. Find the number whose logarithm is 3.36175.
10. Multiply 2261 by 4335.
11. Multiply 6218 by 3998.
12. Multiply 231.9 by 478.8 by 7613 by .921.
13. Multiply .00983 by .0291.
14. Multiply .222 by .00054.
15. Divide 27,683 by 856.
16. Divide 4337 by 38.88.
17. Divide .9286 by 28.75.
18. Divide .0428 by 1.136.
19. Divide 3995 by .003,337.
20. Find the square of 4291.
21. Raise 22.91 to the fourth power.
22. Raise .0236 to the third power.
23. Find the square root of 302,060.
24. Find the cube root of 77.85.
25. Find the square root of .087,64.
26. Find the fifth root of 226,170,000.

1. 3.5879.
2. 1.8667.
3. − 3.9948.
4. 4.6155.
5. 666.2.
6. 74430.
7. .2745.
8. .00024105.
9. 2302.5.
10. 9,802,000.
11. 24,860,000.
12. 778,500,000.
13. .000286.
14. .0001199.
15. 32.34.
16. 111.6.
17. .0323.
18. .03767.
19. 1,198,000.
20. 18,410,000.
21. 275,500.
22. .00001314.
23. 549.7.
24. 4.27.
25. .296.
26. 46.86.

GENERAL PROPERTIES.

IX. LOGARITHMS.

§ 1. GENERAL PROPERTIES.

THE LOGARITHM of a number is the exponent of that power to which another number, the *base*, must be raised to give the number first named. [I. § 11

E.g., in the equation $A^x = N$, A is the base, N the number; and x the exponent of the power of A and the *logarithm to base* A *of the number* N.

The equation $x = \log_A N$ expresses the relation last named.

The equation $N = \log_A^{-1} x$ means that N is the number, A^x, whose logarithm to base A is x; it is read, N is the *anti-logarithm* of x to base A.

E.g., $0 = \log_A 1$ and $A = \log^{-1} 0$, whatever A may be.

So, $\quad 1 = \log_2 2, \quad 2 = \log_3 9, \quad 3 = \log_4 64, \quad 4 = \log_5 625, \cdots,$
and $\quad 2 = \log_2^{-1} 1, \; 9 = \log_3^{-1} 2, \; 64 = \log_4^{-1} 3, \; 625 = \log_5^{-1} 4, \cdots.$

So, $\quad {}^-1 = \log_2 \tfrac{1}{2}, \; {}^-2 = \log_3 \tfrac{1}{9}, \; {}^-3 = \log_4 \tfrac{1}{64}, \; {}^-4 = \log_5 \tfrac{1}{625}, \cdots,$
and $\quad {}^-1 = \log_{\frac{1}{2}} 2, \; {}^-2 = \log_{\frac{1}{3}} 9, \; {}^-3 = \log_{\frac{1}{4}} 64, \; {}^-4 = \log_{\frac{1}{5}} 625 \cdots.$

If the base be well known it may be suppressed, and these two equations may then be written $x = \log N, \; N = \log^{-1} x$.

If while A is constant N take in succession all possible values from 0 to ∞, the corresponding values of x when taken together constitute *a system of logarithms to base* A.

Operations upon or with logarithms are therefore operations upon or with the exponents of the powers of any same base; and the principles established for such powers apply directly to logarithms, with but the change of name noted above.

THEOR. 1. *The logarithm of unity to any base is zero.* [df. pwr.

THEOR. 2. *The logarithm of any number to itself as base is unity.* [df. pwr.

THEOR. 3. *To any positive base* $\begin{cases} larger \\ smaller \end{cases}$ *than unity, every positive number has one and but one real logarithm:*
$\begin{cases} an\ increasing \\ a\ decreasing \end{cases}$ *function of the number.* [VIII. th. 13

LOGARITHMS.

NOTE. If either the base or the number be negative, there may or may not be one real logarithm.

E.g., $^+100$ has the logarithm 2 to base $^+10$ or $^-10$,
and both $^+10$ and $^-10$ have the logarithm $\frac{1}{2}$ to base $^+100$;
but $^-100$ has no real logarithms to base $^+10$ or $^-10$,
nor has $^+10$ or $^-10$ a real logarithm to base $^-100$.

So, $^\pm 1000$ has the logarithm 3 to base $^\pm 10$,
and $^\pm 10$ has the logarithm $\frac{1}{3}$ to base $^\pm 1000$;
but $^\mp 1000$ has no real logarithm to base $^\pm 10$,
and $^\pm 10$ has no real logarithm to base $^\mp 1000$.

In this chapter, and in general where logarithms to the base 10 are used as aids in numerical computations, the number as well as the base is assumed to be positive unless the contrary be stated.

THEOR. 4. *If the base be positive and* $\left\{\begin{matrix}larger\\smaller\end{matrix}\right.$ *than unity, the logarithms of all numbers greater than unity are* $\left\{\begin{matrix}positive\,;\\negative\,;\end{matrix}\right.$ *of all numbers positive and less than unity,* $\left\{\begin{matrix}negative.\\positive.\end{matrix}\right.$ [VIII. lem. th. 12

THEOR. 5. *If the base be positive and* $\left\{\begin{matrix}larger\\smaller\end{matrix}\right.$ *than unity, and if the number be a positive variable that approaches zero, then its logarithm approaches* $\left\{\begin{matrix}negative\ infinity.\\positive\ infinity.\end{matrix}\right.$ [VIII. th. 13

THEOR. 6. *The logarithm of a* $\left\{\begin{matrix}product\\quotient\end{matrix}\right.$ *of two numbers is the* $\left\{\begin{matrix}sum\ of\ the\ logarithms\ of\ the\ factors.\\excess\ of\ log.\ div'd\ over\ log.\ div'r.\end{matrix}\right.$ [VIII. ths. 3, 9

E.g., $\log_A (B \cdot C : D) = \log_A B + \log_A C - \log_A D$.

THEOR. 7. *The logarithm of a* $\left\{\begin{matrix}power\\root\end{matrix}\right.$ *of a number is the* $\left\{\begin{matrix}product\\quotient\end{matrix}\right.$ *of the logarithm of the number by the* $\left\{\begin{matrix}exponent.\\root\text{-}index.\end{matrix}\right.$ [VIII. ths. 4, 10

E.g., $\log_A (B^2 \cdot \sqrt[3]{C}) = 2 \log_A B + \frac{1}{3} \log_A C$.

COR. *The logarithm of the square root of the product of two numbers is the half sum of their logarithms to the same base.* [th. 6

E.g., $\log_A \sqrt{(B \cdot C)} = \frac{1}{2} (\log_A B + \log_A C)$.

[4-8, § 1.]   GENERAL PROPERTIES.

Theor. 8. *If the logarithm of any same number be taken to two different bases, the first logarithm equals the product of the second logarithm into the logarithm of the second base taken to the first base, and* vice versa.

Let N be any number, A, B two bases;
then will $\log_A N = \log_B N \cdot \log_A B$, and $\log_B N = \log_A N \cdot \log_B A$.

For let $y = \log_B N$; [df. log
then $\therefore$ $N = B^y$,
and $\log_A N = y \cdot \log_A B$, [th. 7
$\therefore \log_A N = \log_B N \cdot \log_A B$.   Q.E.D.
So, $\log_B N = \log_A N \cdot \log_B A$.   Q.E.D.

Cor. 1. $\log_A B \cdot \log_B C \cdot \log_C D = \log_A D$.
$\log_A B \cdot \log_B C \cdot \log_C D \cdots \log_K L = \log_A L$.

Cor. 2. *The logarithms of two numbers, each taken to the other number as base, are reciprocals.*

For let $N = A$;
then $\log_A B \cdot \log_B A = \log_A A = 1$.   Q.E.D.

Cor. 3. $\log_A B \cdot \log_B C \cdot \log_C A = 1$;
$\log_A B \cdot \log_B C \cdot \log_C D \cdots \log_K A = 1$.

Note. The reader will observe that the bases and numbers run in cyclic order:

Cor. 4. *The modulus of any system of logarithms is the logarithm, in that system, of the Napierian base* e.   [VIII. th. 15 nt.

Let A be the base of any system of logarithms, and M_A the modulus;
then $\therefore \log_A x = \log_A e \cdot \log_e x$, [th. 8
wherein $\log_A e$ is a constant, independent of x,
$\therefore D_x \log_A x = \log_A e \cdot D_x \log_e x$,
i.e., $\dfrac{M_A}{x} = \log_A e \cdot \dfrac{1}{x}$, [VII. th. 16
$\therefore M_A = \log_A e$.   Q.E.D.

$E.g.$, $M_{10} = \log_{10} e = \log_{10} 2.71828 \cdots$ [
      $= .4342944 \cdots$.

LOGARITHMS. [IX. th. 9

§ 2. SPECIAL PROPERTIES, BASE 10.

The logarithm of an exact power of 10 is an integer. [df. log
E.g., of ···, 1000, 100, 10, 1, .1, .01, .001, ···
the logarithms to base 10 are
···, $+3$, $+2$, $+1$, 0, -1, -2, -3, ···.
But of any other number the logarithm is fractional or incommensurable, and consists of a whole number, the *characteristic*, and an endless decimal, the *mantissa*. [VIII. § 4 df. incom. pwr.

As a matter of convenience the mantissa is always taken positive; and the characteristic is the exponent, positive or negative, of the integral power of 10 next below the given number.

A negative characteristic is indicated by the sign $-$ above it.
E.g., of the numbers
    2000,   20,   .2,   .002,
the logarithms to base 10 are
    3.30103 ···, 1.30103 ···, $\bar{1}$.30103 ···, $\bar{3}$.30103 ···,
whose characteristics are 3, 1, $\bar{1}$, $\bar{3}$,
and whose common mantissa is $+$.30103 ···.

THEOR. 9. *If a number be multiplied (or divided) by any integral power of* 10, *the logarithm of the product (or quotient) and the logarithm of the number have the same mantissa.*

For ∴ the logarithm of a product is the sum of the logarithms
    of its factors.   [th. 6
and ∴ the logarithm of the multiplier is integral,   [hyp.
∴ the mantissa of the sum is identical with the mantissa
    of the logarithm of the multiplicand.   Q.E.D.

So, if a number be divided by an integral power of 10.

COR. *For all numbers that consist of the same significant figures in the same order, the mantissa of the logarithm is constant, but the characteristic changes with the position of the decimal point in the number.*

E.g., of the numbers
    79500,   795,   7.95,   .0795,   .000795,
the logarithms to base 10 are
    4.9004, 2.9004, 0.9004, $\bar{2}$.9004, $\bar{4}$.9004.

§ 3. COMPUTATION OF LOGARITHMS.

PROB. 1. TO COMPUTE THE LOGARITHM OF A NUMBER TO A GIVEN BASE.

FIRST METHOD, BY CONTINUED FRACTIONS.

Form the exponential equation, $A^x = N$, *wherein* N *is the number,* A *the base, and* x *the logarithm sought.* [df. log

By trial find two consecutive integers, x' *and* $x'+1$, *between which* x *lies, and write* $x = x' + y^{-1}$, *wherein* x' *is known and* y^{-1} *is some positive number less than unity.*

In the equation $A^x = N$, *replace* x *by* $x' + y^{-1}$, *giving* $A^{x'+y^{-1}} = N$, *and divide both members by* $A^{x'}$, *giving* $A^{\frac{1}{y}} = N : A^{x'}, \equiv N'$, *say.*

Raise both members of the equation $A^{\frac{1}{y}} = N'$ *to the yth power, giving* $A = N'^y$.

By trial find two consecutive integers, y' *and* $y'+1$, *between which* y *lies, write* $y = y' + z^{-1}$, *and so on, as above.*

Then
$$x = x' + \frac{1}{y} = x' + \cfrac{1}{y' + \cfrac{1}{z}} = x' + \cfrac{1}{y' + \cfrac{1}{z' + \cdots}}$$

and the convergents, which approach x *as their limit, are:*

$$x', \quad \frac{x'y'+1}{y'}, \quad \frac{x'y'z' + z' + x'}{y'z' + 1}, \quad \cdots.$$

E.g., given $10^x = 5$, *to find* x, *i.e. to find* $\log_{10} 5$.

Put $\quad x \quad = 0 + y^{-1}$,

then ∴ $10^{\frac{1}{y}} = 5$, ∴ $5^y = 10$, $y = 1 + \dfrac{1}{z}$;

$5^{1+\frac{1}{z}} = 10$, $5^{\frac{1}{z}} = 2$, $2^z = 5$, $z = 2 + \dfrac{1}{r}$;

$2^{2+\frac{1}{r}} = 5$, $2^{\frac{1}{r}} = \dfrac{5}{4}$, $\left(\dfrac{5}{4}\right)^r = 2$, $r = 3 + \dfrac{1}{s}$;

$\left(\dfrac{5}{4}\right)^{3+\frac{1}{s}} = 2$, $\left(\dfrac{5}{4}\right)^{\frac{1}{s}} = \dfrac{128}{125}$, $\left(\dfrac{128}{125}\right)^s = \dfrac{5}{4}$, $s = 9 + \dfrac{1}{t}$;

and so on.

∴ $x = \cfrac{1}{1+\cfrac{1}{z}} = \cfrac{1}{1+\cfrac{1}{2+\cfrac{1}{r}}} = \cfrac{1}{1+\cfrac{1}{2+\cfrac{1}{3+\cfrac{1}{s}}}} = \cfrac{1}{1+\cfrac{1}{2+\cfrac{1}{3+\cfrac{1}{9+\cdots}}}}$,

and the convergents are

$$1, \quad \frac{2}{3}, \quad \frac{7}{10}, \quad \frac{65}{93}, \quad \cdots.$$

LOGARITHMS. [IX. pr.

These convergents are alternately too large and too small; but their errors are respectively less than
$$\frac{1}{3}; \quad \frac{1}{3\cdot 10}=\frac{1}{30}; \quad \frac{1}{10\cdot 93}=\frac{1}{930}; \quad \frac{1}{93\cdot \text{next denominator}},$$
which denominator is not less than $93+10, =103$; [VI. ths. 1,2

$\therefore \frac{65}{93}, = .69892\cdots$, is too small, and differs from the true value by less than $\frac{1}{9579}$.

The true logarithm of five to seven decimal places, as shown by the tables, is .6989700, so that $\frac{65}{93}$ actually differs from it by less than half of one ten-thousandth.

So, $\log_{10}2 = \log_{10}10 - \log_{10}5 = 1 - .69897 = .30103$.
So, $\log 4 = 2\cdot \log 2 = .60206$; $\log 8 = 3\cdot \log 2 = .90309$;
$\log 625 = 4\cdot \log 5 = 2.79588$; $\log \frac{4}{5} = \log 4 - \log 5 = \bar{1}.90309$.

SECOND METHOD, BY SUCCESSIVE SQUARE ROOTS OF PRODUCTS.

Take two numbers whose logarithms are known, the one greater and the other less than the given number.

Find the square root of their product and the logarithm of this root, the half sum of their logarithms. [th. 10

Multiply this root by whichever of the two numbers lies at the other side of the given number, and find the square root of the product, and the half sum of the logarithms of the factors; and so on.

E.g., to find the logarithm of 5 to the base 10:

Take 10 whose logarithm is 1, and 1 whose logarithm is 0;

	Number.		Logarithm.
then	$\sqrt{(10\times 1)}$	$= 3.16227766$; $\frac{1}{2}(1+0)$	$= .5$
	$\sqrt{(10\times 3.16227766)}$	$= 5.62341325$; $\frac{1}{2}(1+.5)$	$= .75$
	$\sqrt{(3.16228\times 5.62341)}$	$= 4.21696535$; $\frac{1}{2}(.5+.75)$	$= .625$
	$\sqrt{(5.62341\times 4.21697)}$	$= 4.86967671$; $\frac{1}{2}(.75+.625)$	$= .6875$
	$\sqrt{(5.62341\times 4.86968)}$	$= 5.23299218$; $\frac{1}{2}(.75+.6875)$	$= .71875$
	$\sqrt{(4.86968\times 5.23299)}$	$= 5.04806762$; $\frac{1}{2}(.6875+.71875)$	$= .70313$
	$\sqrt{(4.86968\times 5.04807)}$	$= 4.95807276$; $\frac{1}{2}(.6875+.703125)$	$= .69531$
	$\sqrt{(4.95807\times 5.04807)}$	$= 5.00028680$; $\frac{1}{2}(.69531+.70312)$	$= .69921$
	$\sqrt{(4.95807\times 5.00029)}$	$= 4.97709632$; $\frac{1}{2}(.69921+.69531)$	$= .69726$
		$\frac{1}{2}(.69921+.69726)$	$= .69823$

§ 4. TABLES OF LOGARITHMS.

If for successive equidistant values of a variable the corresponding values of a function of this variable be arranged in order, the function is *tabulated;* the variable is the *argument* of the table [I. § 13] and the successive values of the function are the *tabular numbers*. The values of the argument are commonly placed in the margin of the table.

If the logarithms, to any one base, of the successive integers from 1 to a given number, say 1000, or 10000, be arranged for ready reference, they form a *table of logarithms*. Such tables are in use to three places of decimals, to four, five, six, seven, and even ten, twenty, or more places.

In general, the greater the number of decimal places, the greater the accuracy, and the greater the labor of using the tables. For the ordinary use of the engineer, navigator, chemist, or actuary, four- or five-place tables are sufficient; but most refined computations in Astronomy or Geodesy require at least seven-place tables.

Most logarithmic tables are arranged on the same general plan as the four-place table given on pp. 248, 249. This table gives the mantissa only; the computer can readily supply the characteristic. To save space, the first two figures of each argument are printed at the left of the page, and the third figure at the top of the page over the corresponding logarithm.

To save time, labor, and injury to the eyes, the computer should use a well-arranged table, and then train himself to certain habits. The best tables have the numbers grouped by spaces, or by light and heavy lines, into blocks of three or five lines, and three or five columns, corresponding to the right-hand figures of the arguments of the table. The usual patterns are

|0|1 2 3|4 5 6|7 8 9|0|1 2 3| ··· for three-line blocks,
and |0 1 2 3 4|5 6 7 8 9|0 1 2 3 4|··· for five-line blocks,

as in the table on pp. 248, 249. Instead of tracing single lines of figures across the page and down the column, the computer should learn to guide himself by correspondences of position in the blocks.

§ 6. OPERATIONS WITH COMMON LOGARITHMS.

PROB. 2. TO TAKE OUT THE LOGARITHM OF A GIVEN NUMBER.

(a) *One, two, or three significant figures.*

If the number have one significant figure, annex two zeros; if two significant figures, annex one zero; for the mantissa write the four figures that lie opposite the first two figures and under the third figure, and for the characteristic write the exponent of the power of 10 *next below the given number.*

E.g., $\log 567 = 2.7536$; $\log 5.6 = 0.7482$; $\log .05 = \overline{2}.6990$;

If a number have more than three significant figures, the mantissa of its logarithm is not found in the table, but lies between two tabular mantissas whose arguments are two three-figure numbers next larger and next smaller than the given number. [th. 3

E.g., mantissa $\log 500.6$ lies between .6990, .6998,
i.e.,   between mantissa logs 500, 501.

(b) *Four or more significant figures.*

Find the mantissa of the logarithm of the first three figures as above; subtract this mantissa from the next larger tabular mantissa, and take such part of the difference as the remaining figures are of a unit having the rank of the third figure; add this product, as a correction, to the mantissa of the first three figures.

E.g., to find $\log 500.6$;

then ∴ $\log 500 = 2.6990$, $\log 501 = 2.6998$,   [tables
and $\log 501 - \log 500 = .0008$, $500.6 - 500 = .6$,
    ∴ $\log 500.6 = 2.6990 + .6$ of $.0008 = 2.6995$.

NOTE 1. If the given number lie nearer the larger of the two arguments, its mantissa is easiest found by subtracting from the larger of the two tabular mantissas such part of their difference as the excess of the larger argument over the given number is of a unit having the rank of the third figure.

E.g., to find mantissa $\log 500.6$;

then ∴ mantissa logs 500, 501 = .6990, .6998,   [tables
and   ∴ $\log 501 - \log 500 = .0008$, $501 - 500.6 = .4$,
    ∴ mantissa $\log 500.6 = .6998 - .4$ of $.0008 = .6995$.

OPERATIONS WITH COMMON LOGARITHMS.

NOTE 2. The rule for *interpolating* or applying the correction rests upon a property which logarithms have in common with most other functions, and which the reader may observe for himself if he will examine the table carefully, viz.: that the differences of logarithms are very nearly proportional to the differences of their numbers when these differences are small. They are not exactly proportional, but the error is so small as to be inappreciable when using a four-place table. The seven-place tables give the logarithms of all five-figure numbers, and the errors for the sixth, seventh, and eighth figures, as far as due to this cause, are inappreciable. So the rule above given "for applying the correction" is universal.

NOTE 3. The computer should train himself to find the correction and add it to the tabular mantissa (or subtract it) mentally, and to write down only the final result.

To aid in this mental computation, small tables of *proportional parts* are often printed at the side of the principal table. Two forms of such tablets are here shown: the first most accurate, and the other of easiest use.

	19	18			19	18
1	1.9	1.8	or	1	2	2
2	3.8	3.6		2	4	4
3	5.7	5.4		3	6	5
4	7.6	7.2		4	8	7
5	9.5	9.0		5	10	9
6	11.4	10.8		6	11	11
7	13.3	12.6		7	13	13
8	15.2	14.4		8	15	14
9	17.1	16.2		9	17	16

E.g., to find mantissa log 22674;
then ∵ log 227 − log 226 = .3560 − .3541 = .0019,
∴ the correction to be added to .3541 is
.7 of .0019 + .04 of .0019; and is found thus:

opposite 7 find	13.3	or	13	.3541
opposite 4 find	.8		1	+14
Add; the correction is	14		14	giving .3555

Or ∵ 22700 − 22674 = 26,
∴ the correction to be subtracted from .3560 is
.2 of .0019 + .06 of .0019; and is found thus:

opposite 2 find	3.8	or	4	.3560
opposite 6 find	1.1		1	−5
Add; the correction is	5		5	.3555

LOGARITHMS. [IX. Prs.

PROB. 3. TO FIND A NUMBER FROM ITS LOGARITHM.
(a) *The mantissa found in the table.*

Write down the two figures opposite to the given mantissa in the left-hand column, and following them the figure at the top of the column in which the mantissa is found.

Place the decimal point so that the number shall be next above that power of 10 whose exponent is the given characteristic.

E.g., $\log^{-1} 2.7536 = 567$; $\log^{-1} 0.7482 = 5.6$; $\log^{-1} \bar{2}.6990 = .05$.

(b) *The mantissa not found in the table.*

Take out the first three figures for the tabular mantissa next less, as above; from the given mantissa subtract this tabular mantissa, and divide the difference by the difference between the tabular mantissa next less and that next greater.

Annex the quotient to the three figures first found.

Place the decimal point as above.

E.g., to find $\log^{-1} 2.6995$.

then ∴ $\log^{-1} 2.6990 = 500$, $\log^{-1} 2.6998 = 501$, [tables
and ∴ $2.6995 - 2.6990 = .0005$, $2.6998 - 2.6990 = .0008$;
∴ the number sought is $500 + (.0005 : .0008)$, $= 500.6$.

NOTE 1. The process is but the inverse of that for taking out logarithms, and the reason of the rule is the same for both.

This four-place table allows only one-figure corrections, and so gives only four-figure numbers. In general, an n-place table gives n-figure numbers; but sometimes, when the mantissa is large, the nth figure may be two or three units in error, and then the number is approximate only for $n-1$ figures [V. § 5].

NOTE 2. If the given mantissa lie nearer the larger of the two tabular mantissas, the correction may be applied to the larger argument by subtraction.

E.g., to find $\log^{-1} .3555$;

then ∴ the next tabular mantissas .3541, .3560 differ by .0019,
and correspond to 226, 227, as arguments,
and ∴ $.3555 - .3541 = .0014$, $.3560 - .3555 = .0005$,
∴ the number sought is $226 + \frac{14}{19}$, or $227 - \frac{5}{19}, = 226.74$.

3-5, §5.] OPERATIONS WITH COMMON LOGARITHMS.

If the tablets of proportional parts be used, the work, written out, appears as follows:

		or		
14	226		5	227
13.3	+.7		3.8	−.2
.7	+ 4		1.2	− 6
.8	226.74		1.1	226.74,

PROB. 4. TO FIND, BY ONE OPERATION, THE ALGEBRAIC SUM OF SEVERAL LOGARITHMS.

Arrange the logarithms vertically, and take the algebraic sum of each column of digits, beginning at the right and carrying as in ordinary addition; if this sum for any column be negative, make it positive by adding one or more tens to it and subtract as many units from the next column.

E.g., to find the algebraic sum in the margin, 3.1037
adding upward, the computer says: − 0.6986

9, 7, 16, 10, 17, $+ \bar{2}.2409$
1, 3, ⁻6, ⁻14, ⁻11, 9,   2 off, $- \bar{2}.5892$
⁻2, 3, ⁻5, ⁻1, ⁻10, 0,   1 off, $+ \bar{1}.2529$
⁻1, 1, ⁻4, ⁻2, ⁻8, ⁻7, 3, 1 off, $= 1.3097$
⁻1, ⁻2, 0, ⁻2, 1,

and,   adding downward, for a check, he says:

7, 1, 10,   8, 17,
1, 4, ⁻4, ⁻13, ⁻11, 9,   2 off, and so on.

PROB. 5. TO DIVIDE A LOGARITHM WHOSE CHARACTERISTIC IS NEGATIVE.

Write down the number of times the divisor goes into that multiple of itself which is equal to, or next less than, the negative characteristic; carry on the positive remainder to the mantissa, and divide.

E.g., $\bar{4}.1234 : 3 = (^-6 + 2.1234) : 3 = \bar{2}.7078.$

So, $\bar{3}.4770 \cdot \tfrac{1}{2} = \bar{8}.4310 : 2 = \bar{4}.2155.$

LOGARITHMS. [IX. pra.

PROB. 6. TO AVOID NEGATIVE CHARACTERISTICS.

Modify the logarithms by adding 10 *to their characteristics when negative; use the sums, differences, or exact multiples of the modified logarithms where the subject-matter is such that the computer cannot mistake the general magnitude of the results.*

To divide a modified logarithm, add such a multiple of 10 *as will make the modified logarithm exceed the true logarithm by* 10 *times the divisor; then divide.*

E.g., if $\log a = \bar{2}.3010$, $\log b = \bar{1}.4771$, to find $\log(a^{\frac{2}{3}} b^{-\frac{3}{5}})$,
$= \frac{1}{5}(2 \log a - 3 \log b)$.

BY TRUE LOGARITHMS.	BY MODIFIED LOGARITHMS.
$\bar{2}.3010 \cdot 2 = \bar{4}.6020$	$8.3010 \cdot 2 = 6.6020$
$\bar{1}.4771 \cdot 3 = \bar{2}.4313$	$9.4771 \cdot 3 = 8.4313$
$5)\overline{\bar{2}.1707}$	$5)\overline{8.1707}$
$\bar{1}.6341$	9.6341

At each step of the work with modified logarithms, any tens in the characteristics are rejected, or tens, if necessary, are added, so as to keep the characteristics between 0 and 9 inclusive. Before dividing by 5, in the example just above, 4 tens were added, making the dividend 48.1707.

NOTE. The *arithmetical complement* of the logarithm of a number is the modified logarithm of the reciprocal of the number. It is got by subtracting the given logarithm, modified, if necessary, from 10; it may be read from the table by subtracting each figure from 9, beginning with the characteristic and ending with the last significant figure but one, subtracting the last significant figure from 10, and annexing as many zeros as the given logarithm ends with. The arithmetical complement of the arithmetical complement is the original logarithm.

E.g., ar-com 3.4908000 = 6.5092000, and conversely.

In any algebraic sum, a subtractive logarithm can be replaced by its arithmetical complement taken additively. In most cases, however, the method of prob. 4 appears preferable.

E.g., in the example under prob. 4, the terms -0.6986, $-\bar{2}.5892$ might be replaced by 9.3014, 1.4108.

OPERATIONS WITH COMMON LOGARITHMS.

Prob. 7. To compute by logarithms the products, quotients, powers, and roots of numbers.

1. *For a product:* add the logarithms of the factors, and take out the antilogarithm of the sum.

2. *For a quotient:* from the logarithm of the dividend subtract that of the divisor, and take out the antilogarithm.

3. *For a power:* multiply the logarithm of the base by the exponent of the power sought, and take out the antilogarithm.

4. *For a root:* divide the logarithm of the base by the root-index, and take out the antilogarithm.

E.g., to find the value of $(.01519 \cdot 6.318 : 7.254)^{\frac{2}{3}}$:

NUMBERS.	LOGARITHMS.
.01519	$\bar{2}.1815$
× 6.318	+0.8006
÷ 7.254	−0.8605
	$\bar{2}.1216 \times \frac{2}{3}$
and the number sought is 0.001522.	$\bar{3}.1824$

Note. Not only simple operations, as in the above example, but complex operations, can be performed by logarithms. Sometimes the expression whose value is sought must first be prepared by factoring.

E.g., to find the value of $\sqrt{(h^2 - b^2)}$, wherein h, b are any given numbers and may represent the lengths of the hypothenuse and base of a right triangle:

then $\quad \sqrt{(h^2 - b^2)} = \log^{-1} \frac{1}{2} (\log \overline{h + b} + \log \overline{h - b})$.

Prob. 8. To solve the exponential equation $A^x = B$.

Divide the logarithm of B by the logarithm of the base A of the exponential: the quotient is x, the exponent sought.

For $\quad \because A^x = B,$
$\quad \therefore x \log A = \log B,$
$\quad \therefore x = \log B : \log A.$ $\qquad$ Q.E.D.

PROB. 9. TO ESTIMATE THE AMOUNT OF POSSIBLE ERROR IN A LOGARITHM OR ANTILOGARITHM GOT FROM THE TABLE, AND IN THE SOLUTIONS OF PROBS. 7, 8:

Let p be the number of decimal places in the table used; $A', B', \cdots X', (A^m B^n \cdots)'$, the number of units of their last decimal places contained in $A, B, \cdots X, A^m B^n \cdots$; $a, \beta, \cdots$, the possible relative errors, all taken positive, of $A, B, \cdots$: then

(a) Poss. err. $\log x = 10^{-p} + .43$ poss. rel. err. x.

(b) Poss. rel. err. $x = 1 : 2 x' + 2.3 \cdot$ poss. err. $\log x$.

(c) Poss. rel. err. $A^m B^n \cdots$ [in pr. 7]
$$= 1 : 2 (A^m B^n \cdots)' + 2.3 (^+m + ^+n + \cdots) \cdot 10^{-p}$$
$$+ (^+m a + ^+n \beta + \cdots).$$

(d) Poss. rel. err. x [in pr. 8]
$$= \frac{1}{2 x'} + \frac{10^{-p} + .43 a}{\log A} + \frac{10^{-p} + .43 \beta}{\log B}.$$

For $\because D_x \log_{10} x = M_{10} \frac{1}{x},$ [VIII. th. 15, $A = 10$

$\therefore M_{10} = x \cdot D_x \log_{10} x \doteq \frac{x \cdot \text{inc} \log x}{\text{inc } x} = .43,$ [table logs

$\therefore \frac{\text{inc} \log x}{\text{inc } x} = \frac{.43}{x},$

i.e., $\text{inc} \log x \doteq .43 \frac{\text{inc } x}{x},$

and $\frac{\text{inc } x}{x} \doteq \frac{\text{inc} \log x}{.43} = 2.3 \text{ inc. } \log x :$

(a) $\because$ $\log x$, as got from x by p-place logarithm-tables, has a possible error composed of:

two possible half-units in pth decimal place, from the omitted decimals of the printed logarithm and of the correction for interpolation,

and   an increment or error, $\doteq .43 \cdot \frac{\text{inc. or err. of } x}{x}$; [above

$\therefore$ poss. err. $\log x \doteq (\frac{1}{2} + \frac{1}{2}) 10^{-p} + .43 \cdot \frac{\text{err. } x}{x}$

$= 10^{-p} + .43 \cdot$ poss. rel. err. x.   Q.E.D.

(b) $\because$ x, as got from $\log x$ by the same table, has a possible error composed of:

9, § 5.] OPERATIONS WITH COMMON LOGARITHMS.

a possible half-unit in last decimal place, for the omitted decimals,

and   an increment or error, $\doteq 2.3 \cdot x \cdot$ inc. log x,

∴ poss. err. $x \doteq \frac{1}{2}$ in last decimal place of $x + 2.3 \cdot x \cdot$ poss. err. log x,

∴ poss. rel. err. $x \doteq 1 : 2x' + 2.3 \cdot$ poss. err. log x.   Q.E.D.

(c)· ∵ poss. err. log A $= 10^{-p} + .43\,a$,   [(a)

∴ poss. err. log $A^m = m\,(10^{-p} + .43\,a$

So,   poss. err. log $B^n = n\,(10^{-p} + .43\,\beta)$, ⋯,

∴ poss. err. log $(A^m B^n \cdots)$
$= (^+m +^+n +\cdots)\,10^{-p} + .43(^+m a +^+n\beta +\cdots)$,

∴ poss. rel. err. $(A^m B^n \cdots)$
$= 1 : 2x' + 2.3[(^+m +^+n +\cdots)10^{-p} + .43(^+m a +\cdots)]$,

wherein $x \equiv A^m B^n \cdots$;

∴ poss. rel. err. $(A^m B^n \cdots)$
$= 1 : 2x' + 2.3(^+m +^+n +\cdots)\,10^{-p} + (^+m a +\cdots)$.
                                                Q.E.D.

(d)   ∵ $x = \log B : \log A$,

∴ poss. rel. err. $x =$ poss. rel. err. from omitted decimals of x
+ poss. rel. err. log A + poss. rel. err. log B   [V. th. 5 cr. 3

$= \dfrac{1}{2x'} + \dfrac{10^{-p} + .43\,a}{\log A} + \dfrac{10^{-p} + .43\,\beta}{\log B}$.   Q.E.D. [(a)

NOTE. If in (d) the divisions log B : log A be performed by logarithms,

then ∵ log x = log log B − log log A,

∴ poss. err. log x = poss. err. log·log A + poss. err. log·log B
$= 10^{-p} + .43$ poss. rel. err. log A
$+ 10^{-p} + .43$ poss. rel. err. log B   [(a)
$= 2 \cdot 10^{-p} + .43\left(\dfrac{10^{-p} + .43\,a}{\log A} + \dfrac{10^{-p} + .43\,\beta}{\log B}\right)$,

∴ poss. rel. err. $x, = \dfrac{1}{2x'} + 2.3$ poss. err. log x   [(b)

$= \dfrac{1}{2x'} + 4.6 \cdot 10^{-p} + \dfrac{10^{-p} + .43\,a}{\log A} + \dfrac{10^{-p} + .43\,\beta}{\log B}$,

which differs from the former result only by the term $4.6 \cdot 10^{-p}$ arising from the omitted decimals of the table used in performing the division, and obtainable also from (c) by making $m = n = 1$, $a = \beta = 0$.

LOGARITHMS.

N	0	1	2	3	4	5	6	7	8	9
10	0000	0043	0086	0128	0170	0212	0253	0294	0334	0374
11	0414	0453	0492	0531	0569	0607	0645	0682	0719	0755
12	0792	0828	0864	0899	0934	0969	1004	1038	1072	1106
13	1139	1173	1206	1239	1271	1303	1335	1367	1399	1430
14	1461	1492	1523	1553	1584	1614	1644	1673	1703	1732
15	1761	1790	1818	1847	1875	1903	1931	1959	1987	2014
16	2041	2068	2095	2122	2148	2175	2201	2227	2253	2279
17	2304	2330	2355	2380	2405	2430	2455	2480	2504	2529
18	2553	2577	2601	2625	2648	2672	2695	2718	2742	2765
19	2788	2810	2833	2856	2878	2900	2923	2945	2967	2989
20	3010	3032	3054	3075	3096	3118	3139	3160	3181	3201
21	3222	3243	3263	3284	3304	3324	3345	3365	3385	3404
22	3424	3444	3464	3483	3502	3522	3541	3560	3579	3598
23	3617	3636	3655	3674	3692	3711	3729	3747	3766	3784
24	3802	3820	3838	3856	3874	3892	3909	3927	3945	3962
25	3979	3997	4014	4031	4048	4065	4082	4099	4116	4133
26	4150	4166	4183	4200	4216	4232	4249	4265	4281	4298
27	4314	4330	4346	4362	4378	4393	4409	4425	4440	4456
28	4472	4487	4502	4518	4533	4548	4564	4579	4594	4609
29	4624	4639	4654	4669	4683	4698	4713	4728	4742	4757
30	4771	4786	4800	4814	4829	4843	4857	4871	4886	4900
31	4914	4928	4942	4955	4969	4983	4997	5011	5024	5038
32	5051	5065	5079	5092	5105	5119	5132	5145	5159	5172
33	5185	5198	5211	5224	5237	5250	5263	5276	5289	5302
34	5315	5328	5340	5353	5366	5378	5391	5403	5416	5428
35	5441	5453	5465	5478	5490	5502	5514	5527	5539	5551
36	5563	5575	5587	5599	5611	5623	5635	5647	5658	5670
37	5682	5694	5705	5717	5729	5740	5752	5763	5775	5786
38	5798	5809	5821	5832	5843	5855	5866	5877	5888	5899
39	5911	5922	5933	5944	5955	5966	5977	5988	5999	6010
40	6021	6031	6042	6053	6064	6075	6085	6096	6107	6117
41	6128	6138	6149	6160	6170	6180	6191	6201	6212	6222
42	6232	6243	6253	6263	6274	6284	6294	6304	6314	6325
43	6335	6345	6355	6365	6375	6385	6395	6405	6415	6425
44	6435	6444	6454	6464	6474	6484	6493	6503	6513	6522
45	6532	6542	6551	6561	6571	6580	6590	6599	6609	6618
46	6628	6637	6646	6656	6665	6675	6684	6693	6702	6712
47	6721	6730	6739	6749	6758	6767	6776	6785	6794	6803
48	6812	6821	6830	6839	6848	6857	6866	6875	6884	6893
49	6902	6911	6920	6928	6937	6946	6955	6964	6972	6981
50	6990	6998	7007	7016	7024	7033	7042	7050	7059	7067
51	7076	7084	7093	7101	7110	7118	7126	7135	7143	7152
52	7160	7168	7177	7185	7193	7202	7210	7218	7226	7235
53	7243	7251	7259	7267	7275	7284	7292	7300	7308	7316
54	7324	7332	7340	7348	7356	7364	7372	7380	7388	7396

LOGARITHMS.

N	0	1	2	3	4	5	6	7	8	9
55	7404	7412	7419	7427	7435	7443	7451	7459	7466	7474
56	7482	7490	7497	7505	7513	7520	7528	7536	7543	7551
57	7559	7566	7574	7582	7589	7597	7604	7612	7619	7627
58	7634	7642	7649	7657	7664	7672	7679	7686	7694	7701
59	7709	7716	7723	7731	7738	7745	7752	7760	7767	7774
60	7782	7789	7796	7803	7810	7818	7825	7832	7839	7846
61	7853	7860	7868	7875	7882	7889	7896	7903	7910	7917
62	7924	7931	7938	7945	7952	7959	7966	7973	7980	7987
63	7993	8000	8007	8014	8021	8028	8035	8041	8048	8055
64	8062	8069	8075	8082	8089	8096	8102	8109	8116	8122
65	8129	8136	8142	8149	8156	8162	8169	8176	8182	8189
66	8195	8202	8209	8215	8222	8228	8235	8241	8248	8254
67	8261	8267	8274	8280	8287	8293	8299	8306	8312	8319
68	8325	8331	8338	8344	8351	8357	8363	8370	8376	8382
69	8388	8395	8401	8407	8414	8420	8426	8432	8439	8445
70	8451	8457	8463	8470	8476	8482	8488	8494	8500	8506
71	8513	8519	8525	8531	8537	8543	8549	8555	8561	8567
72	8573	8579	8585	8591	8597	8603	8609	8615	8621	8627
73	8633	8639	8645	8651	8657	8663	8669	8675	8681	8686
74	8692	8698	8704	8710	8716	8722	8727	8733	8739	8745
75	8751	8756	8762	8768	8774	8779	8785	8791	8797	8802
76	8808	8814	8820	8825	8831	8837	8842	8848	8854	8859
77	8865	8871	8876	8882	8887	8893	8899	8904	8910	8915
78	8921	8927	8932	8938	8943	8949	8954	8960	8965	8971
79	8976	8982	8987	8993	8998	9004	9009	9015	9020	9025
80	9031	9036	9042	9047	9053	9058	9063	9069	9074	9079
81	9085	9090	9096	9101	9106	9112	9117	9122	9128	9133
82	9138	9143	9149	9154	9159	9165	9170	9175	9180	9186
83	9191	9196	9201	9206	9212	9217	9222	9227	9232	9238
84	9243	9248	9253	9258	9263	9269	9274	9279	9284	9289
85	9294	9299	9304	9309	9315	9320	9325	9330	9335	9340
86	9345	9350	9355	9360	9365	9370	9375	9380	9385	9390
87	9395	9400	9405	9410	9415	9420	9425	9430	9435	9440
88	9445	9450	9455	9460	9465	9469	9474	9479	9484	9489
89	9494	9499	9504	9509	9513	9518	9523	9528	9533	9538
90	9542	9547	9552	9557	9562	9566	9571	9576	9581	9586
91	9590	9595	9600	9605	9609	9614	9619	9624	9628	9633
92	9638	9643	9647	9652	9657	9661	9666	9671	9675	9680
93	9685	9689	9694	9699	9703	9708	9713	9717	9722	9727
94	9731	9736	9741	9745	9750	9754	9759	9763	9768	9773
95	9777	9782	9786	9791	9795	9800	9805	9809	9814	9818
96	9823	9827	9832	9836	9841	9845	9850	9854	9859	9863
97	9868	9872	9877	9881	9886	9890	9894	9899	9903	9908
98	9912	9917	9921	9926	9930	9934	9939	9943	9948	9952
99	9956	9961	9965	9969	9974	9978	9983	9987	9991	9996

§ 6. EXAMPLES.

1. What is the logarithm of 144:
   to base $2\sqrt{3}$? to base $2\sqrt[3]{12}$? to base $(2\sqrt[3]{12})^{-1}$?
2. What is the characteristic of:
   $\log_2 7$? $\log_7 2$? $\log_3 21$? $\log_{21} 3$? $\log_{\frac{1}{3}} 21$? $\log_{\frac{1}{3}} 21^{-1}$?
3. Find $\log_5 3125$; $\log_7 343^{-1}$; $\log_{\frac{1}{3}} 81$; $\log_{\frac{1}{3}} 343$; $\log_{\frac{1}{7}} 343^{-1}$.

§ 3, PROB. 1.

4. By continued fractions derive the logarithms, to base 10, of 3 and 7 to four decimal places.
   Thence find the logarithms of:
   9, 2.7, .81, 70, 4.9, 343, 21, 63, .441, $.7^{-1}$, 18.9^{-1}.

§ 5, PROB. 2.

5. From the table take out the logarithms of:
   12, 120, 123, 124, 123.4, 1.234, 12350, .001235.

§ 5, PROB. 3.

6. From the table find the antilogarithms of:
   1.0792, 2.0792, 2.0899, 2.0934, 2.0913, 0.0913, $\bar{4}.0917$.

§ 5, PROBS. 4–8.

7. By logarithms find the values of:
   $$\frac{2^2 5^3 85^2}{3^2 7^3}, \quad \frac{\sqrt{(97^2 - 9^2)}}{81 \cdot \sqrt[3]{572}}, \quad \frac{\sqrt{12} \cdot \sqrt[3]{65}}{\sqrt{5} \cdot \sqrt[7]{.18}}, \quad \frac{\sqrt[3]{83.64} \times 39.56^2}{.08145^2 \times \sqrt[5]{1.968}}$$

8. From the logarithm of 2 find the number of digits in:
   2^{64}, 2^{500}, 5^{100}, 20^8, 160^{10}, 25^{25}, $6.25^{6.25}$, 25^{-4}, 50^{-50}.

9. By logarithms multiply 575.25 by 1.06^{20}; by 1.03^{40}; by 1.015^{80}.
10. By logarithms find $\sqrt[3]{1000}$, $\sqrt[5]{.00010098}$, $\sqrt[7]{.0000000037591}$.
11. What power is 2 of 1.05? 3 of 1.04? 4 of 1.03? 5 of 1.02?
12. If the number of births per year be 1 in 45, and of deaths 1 in 60, in how many years will the population double, taking no account of other sources of increase or decrease?

§ 5, PROB. 9.

13. Find the possible error in each of the examples in Nos. 7–12.

Logarithms.

257.

Before logarithms were discovered and complete tables of logarithms computed, practical arithmetic may be said to have come to a standstill. To-day, calculations, which in the time of Kepler required days and weeks for their completion, or were, to the great loss of mankind, abandoned because of the immense labor involved in their computation; to-day, such calculations may be made in a few minutes by the use of logarithms, even by the veriest beginner in mathematics. It has well been said that logarithms are to arithmetic what the steam engine is to mechanics.

In order to obtain a clear notion of the principle on which logarithms are based, let us take any number, such as 2, for example, and raise it to a series of powers beginning from 0, thus:

$$1 = 2^0, \quad 2 = 2^1, \quad 4 = 2^2, \quad 8 = 2^3, \quad 16 = 2^4, \text{ etc.}$$

Then in order better to examine the relations which exist between these powers and their exponents, let us write them in vertical columns with a dividing line between. The base number 2 and the sign of equality we may omit as being the same for all of the powers here given. Thus instead of $1 = 2^0$, $128 = 2^7$, we shall write more briefly, $1 \mid 0$, $128 \mid 7$, etc.

Powers	Exponents	Powers	Exponents	Powers	Exponents
1	0	1024	10	1048576	20
2	1	2048	11	2097152	21
4	2	4096	12	4194304	22
8	3	8192	13	8388608	23
16	4	16384	14	16777216	24
32	5	32768	15		
64	6	65536	16		
128	7	131072	17		
256	8	262144	18		
512	9	524288	19		

258.

Now remembering the four fundamental rules for powers, (§ 204 to 208),

$$a^m \cdot a^n = a^{m+n} \qquad (1)$$

$$\frac{a^m}{a^n} = a^{m-n} \qquad (2)$$

$$(a^n)^m = a^{mn} \qquad (3)$$

$$\sqrt[n]{a^m} = a^{\frac{m}{n}} \qquad (4)$$

we see how such a table of powers and exponents can be made of use.

With the exception of addition and subtraction, any operations desired upon any numbers in the first column can be greatly abridged if we take instead the corresponding exponents, and thus convert multiplication into addition, division into subtraction, the raising of powers into multiplication and the extraction of roots into simple division. All these can be accomplished by such a simple, but invaluable table of powers and exponents.

EXAMPLE 1. If two or more numbers, such as 128 and 512 are to be multiplied together, we look for these factors in the first column, and take the corresponding exponents and *add* them together; then find in the column of exponents one which is equal to their sum and opposite it will be the desired product. Thus we have it in our little table:

$$\begin{array}{r} \text{Exponent of } 128 = 7 \\ \text{``}\quad\text{``}\quad 512 = 9 \\ \hline 16 \end{array}$$

Opposite the exponent 16 we find the power 65536, which is the product of the two factors.

The reason for this is readily seen. All the numbers which are in the first columns are powers of the same base number. Thus in the above example:
$$128 = 2^7$$
$$512 = 2^9$$

But $128 \cdot 512 = 2^7 \cdot 2^9 = 2^{16}$ (see § 204), hence that power of 2 which has the exponent 16 must be the product of the given numbers, or
$$128 \cdot 512 = 65536.$$

EXAMPLE 2. Suppose it is required to divide any of the numbers in the first column by any other of these, we merely *subtract* the exponent of the divisor from the exponent of the dividend, and the power which corresponds to the resulting difference, will be the quotient.

Thus we have:
$$\frac{2097152}{256} = 8192$$

Because:

Exponent of $2097152 = 21$
"   " $256 = 8$

Difference   13

$8192 =$ power whose exponent is $= 13$.

This is the same as:
$$\frac{2097152}{256} = \frac{2^{21}}{2^8} = 2^{21-8} = 2^{13} = 8192 \qquad \text{(see § 205)}$$

EXAMPLE 3. If it is required to raise any quantity in the first column to any given power, we simply *multiply* the exponent in the table by that of the required power, and opposite the product we find the required power. Thus:
$$16^5 = 1048576$$

We have: Exponent of $16 = 4$
Multiplied by $\underline{5}$
20

And the power whose exponent in the table is 20, is 1048576.

For $16 = 2^4$
whence $16^5 = (2^4)^5 = 2^{20} = 1048576$ (see § 207)

EXAMPLE 4. If it is required to extract the root of any number in the first column, we simply divide the exponent of the given number by the exponent of the required root, and the power corresponding to the quotient is the required root. Thus:

$$\sqrt[7]{2097152} = 8$$

This follows because :

Exponent of $2097152 = 21$

$$\frac{21}{7} = 3$$

Power whose exponent is 3, is 8,

or $2097152 = 2^{21}$

$$\sqrt[7]{2097152} = \sqrt[7]{2^{21}} = 2^{\frac{21}{7}} = 2^3 \quad \text{(See § 208)}$$

259.

We see the great value even of such an imcomplete table as that given in the preceding paragraph, but its usefulness is very limited because of the great gaps which exist in it. If we desire to make calculations involving numbers between 2 and 4, 4 and 8, 8 and 16, etc., we are unable to use the table, because it does not contain these numbers.

When, however, these numbers are all introduced, and their corresponding exponents calculated, then the system becomes complete and is of the highest degree of utility for all calculations. Such a table is called a Table of Logarithms. The quantities in the first column are called the *Numbers,* while their corresponding exponents in the second column are called the *Logarithms.* We have also to discuss in this connection the base number upon which the exponents are calculated, and this will be referred to as the *Base.*

Notwithstanding the great value of such tables of logarithms, no private individual could take the time and labor to calculate a table for himself. This gigantic work has, however, been very fully and accurately done already, and many excellent editions of logarithms published.

The best tables are those calculated to seven decimal places, and the best of these are the magnificent foreign editions of Schrön and of Bruhns. The American edition of Stanley is also good. For many purposes six decimals are sufficient, and good six figure tables are those of Loomis, published by Harper & Bros., and those published by Prof. Geo. W. Jones at Ithaca, N. Y. Very often five decimals are quite enough for ordinary calculations and the five figure logarithms contained in Trautwines Tables, will be found very useful.

Logarithms were invented in the early part of the 17th century by Baron Napier, of Merchiston, of Scotland. The first full table of logarithms was calculated shortly afterward by Henry Briggs, also of Scotland. the work requiring an entire year and the help of eight assistants. This table was to 14 decimal places, and for all numbers up to 10,000. It was afterwards extended to all numbers up to 100,000, by Adrian Vlacq, a Hollander. Vlacq's tables were to 10 decimal places, and on these original tables, with many errors corrected, the modern extremely accurate tables are all based.

260.

It is evidently immaterial, so far as the principle is concerned, what number is taken as the *base* for a system of logarithms. When the system is once made, the base is no longer used. In the little table in § 257 the base is 2. Any other number may be taken, however, and the resulting system will be equally useful. Since the base may be chosen at will, all the tables above referred to, derived from the original tables of Briggs, have been calculated on the same base as our number system, i. e., 10.

Logarithms calculated to the base 10, are called *common* logarithms, or sometimes, Briggs' logarithms. There is also another, the so-called Natural System (also called Hyperbolic Logarithms), which, although not so convenient in practice is of the greatest importance in higher mathematics, and in steam engineering.

In the following pages it is assumed that the student has provided himself with a good set of logarithmic tables, and as the following problems are all worked with seven-figure logarithms, he had better procure a good set, such as Schrön or Bruhns, although the cheaper sets will answer if these are not at hand. For the student of engineering or physical science it is very desirable that he should provide himself with a good set of logarithms and acquire every possible facility in their use, and the familiarity which one obtains with the pages of a special, personal copy is such as to make it desirable that he should get at the beginning the set he intends to keep as his companion in all his subsequent scientific work.

In nearly all sets of logarithmic tables instructions and explanations for their use are also given, but as these are not always expressed as clearly as might be desired, we will here proceed to explain them in a full and practical manner. No explanations, however, can take the place of frequent practical use, and the greatest facility in the use of such tables must be obtained by actual practice.

261.

In examining a set of seven-figure common logarithms we find among the various values the following, which will serve as an explanation:

NUMBER	LOGARITHM	NUMBER	LOGARITHM
1	0.0000000	10	1.0000000
2	0.3010300		
3	0.4771213	99	1.9956352
4	0.6020600	100	2.0000000
5	0.6989700		
6	0.7781513	999	2.9995655
		1000	3.0000000
		9999	3.9999566

According to § 257 these numbers simply mean that $1 = 10^0$;

$$2 = 10^{0.301030} = 10^{\frac{301030}{1000000}} = \sqrt[1000000]{10^{301030}}$$

$3 = 10^{0.4771213}$, etc.; $10 = 10^1$; $100 = 10^2$, etc.

The Briggs logarithms are therefore nothing more than the *exponents* of those powers to which the number 10 must be raised in order to produce the corresponding *numbers* opposite them in the table. The sign of equality, and the base 10, are omitted from the tables to save room, and are always understood. We see that 0 is the logarithm of 1, since the only power of 10, which is equal to 1, is the 0th power; the logarithm of 2 is 0.3010300, that being the power to which 10 must be raised, to equal 2; the logarithm of 10 is 1, since the 1st power of 10=10.

Therefore we may write:

log. 1=0.0000000; log. 10=1.0000000; log. 2=0.3010300; log. 99=1.9956352, etc.

262.

In the higher Analysis, short methods are derived by means of which a skilled computer can calculate logarithms very rapidly.

Since the student is not yet prepared to study those methods, we shall here explain the more tedious arithmetical method by which they may be calculated. This method is the one by which Briggs calculated his tables, the modern system of higher mathematics not having been invented in his time.

In order to calculate the logarithm of any number, such as 5, for example, we proceed as follows:

The number 5, considered as a power of 10, lies somewhere between 10^0, and 10^1, since $10^0 < 5$, and $10^1 > 5$.

Hence we think that it *may* be equal to 10 raised to the $\frac{0+1}{2} = \frac{1}{2}$ power. We find by trial that $10^{\frac{1}{2}} = \sqrt{10} = 3.1622776601$. This is too small, for $10^{\frac{1}{2}} < 5$; so the logarithm must lie between the narrower limits of $\frac{1}{2}$ and 1, since $10^{\frac{1}{2}} < 5$, and $10^1 > 5$. In this manner we can reduce the limits narrower and narrower, by taking repeatedly the half sum of the greater and lesser exponents, and yet never have to extract a higher root than the square root. Thus, suppose now we try the $\frac{\frac{1}{2}+1}{2} = \frac{3}{4}$ power of 10. We have (§ 210):

$$10^{\frac{3}{4}} = \sqrt{10^{\frac{3}{2}}} = \sqrt{(10^1 \cdot 10^{\frac{1}{2}})} = \sqrt{10 \cdot (3.1622776601)} = 5.6234132$$

hence, $10^{\frac{1}{2}} < 5$, and $10^{\frac{3}{4}} > 5$

Again:

$$10^{\frac{\frac{1}{2}+\frac{3}{4}}{2}} = 10^{\frac{5}{8}} = \sqrt{10^{\frac{5}{4}}} = \sqrt{(10^{\frac{1}{2}} \cdot 10^{\frac{1}{2}})} =$$
$$= \sqrt{(5.623\ldots)(3.1622\ldots)} = 4.21695034$$

hence $10^{\frac{5}{8}} < 5$, and $10^{\frac{3}{4}} > 5$

Again:
$$10^{\frac{5+3}{2}} = 10^{\frac{11}{16}}\sqrt{(10^{\frac{5}{8}} \cdot 10^{\frac{3}{4}})} = \sqrt{(4.2169..)} \ (5.623..) = 4.869675252$$

hence, $10^{\frac{11}{16}} < 5$, and $10^{\frac{3}{4}} > 5$

Again:
$$10^{\frac{11+3}{2}} = \sqrt{(10^{\frac{11}{16}} \cdot 10^{\frac{3}{4}})} = 5.232991, \text{ etc.}$$

We thus see that the approximation is growing constantly nearer and nearer and after repeating the operation 22 times we find that: $10^{\frac{2931088}{4194304}} = 5.00000086$, or $= 5$ correct to the sixth decimal, and hence log. $5 = \frac{2931088}{4194304} = 0.6989700$, it being much more convenient to have the logarithm expressed as a decimal.

Logarithms cannot be calculated with absolute exactness, because they belong to the so-called irrational quantities, of which the decimals are endless.

263.

If the original tables of Briggs and Vlacq had not already been calculated by the above tedious process, the work might have been greatly reduced by means of modern methods. It is only necessary in this way first to compute the logarithms of the prime numbers, since all the others can be derived from them by simple addition and multiplication. This follows, because all composite numbers can be resolved into their prime factors. If we know the logarithms of the various factors of a number, we have simply to add the logarithms of the factors together to obtain the logarithm of the product. Thus if we add together the logarithms of 2 and 3, we obtain the logarithm of 6. The reason for this is readily seen if we call log. $2 = a$, and log. $3 = b$, whence $2 = 10^a$ and $3 = 10^b$, whence $10^a \cdot 10^b = 10^{a+b}$.

By comparison with the table of page 257 we see how the addition of the logarithms of 2 and of 3 gives the logarithm of 6, and we must observe carefully the following points.

In the calculation of the original tables, more than seven decimal places were computed. Of these the first seven were retained, in the preparation of seven figure tables, and the last decimal was increased by 1 whenever the 8th decimal was greater than 5. For this reason exact dependence cannot be placed upon the accuracy of the last decimal. This variation from exactness can, however, be neglected to a great extent in actual practice, since the influence of the last decimal of the logarithm has such a slight influence upon the result as to be of little importance. If we have many logarithms to add together, or have to multiply a logarithm by a large number, the result will come out too large if the last figure is too large. But if, as is often the case, we have again to divide this resulting logarithm by another number, the error will again be reduced. In some tables, such as Schrön's, there is a mark placed under the last decimal whenever it has been so increased, and by taking this into account a correction can be made when extreme accuracy is necessary.

264.

In the Briggs system of logarithms, of which the base number is 10, we have the logarithm of $10=1$, and the logarithm of $1=0$, hence for all numbers between 1 and 10, the logarithms must be greater than 0 and less than 1, i. e., must be proper fractions. Also, since the logarithm of 100 is 2, the logarithms of all numbers between 10 and 100 must be >1 and <2, and must be mixed numbers, or consist of a whole number and a decimal. That portion of a logarithm which consists of a whole number, is called the *characteristic*, while the decimal part is called the *mantissa*, and as these names will frequently be used hereafter, care should be taken to remember them. In the common system, the logarithms of all numbers except those which are even powers of 10, are mixed numbers, and *the characteristic of any logarithm is always 1 less than there are figures in the number.*

For all numbers of one figure, the characteristic is 0, for all numbers of two figures the characteristic is 1, for all 3 figure numbers it is 2, etc., etc. Since the characteristic of the logarithm of any number can thus always be obtained at once by inspection of the number, it is unnecessary to have the characteristics in the tables; and they are therefore omitted in order to save room. In looking in a table, therefore, such as Schrön's or Bruhn's, we find only the decimal part of the logarithm, and must prefix the characteristic ourselves by the above rule. For example:

log. $4571 = 3.6600112$
log. $4577 = 3.6605809$, etc.

265.

Since by adding together the logarithms of several factors we obtain the logarithm of their product, and since the logarithms of all the simple units (10, 100, 1000, etc.), have characteristics which are whole numbers, and mantissas=0, i. e., log. 1=0, log. 10=1, log. 100=2, etc., it is clear that when any number is multiplied by 10, 100, 1000, etc., the mantissa of its logarithm remains unchanged, and only the characteristic is altered.

If, therefore, we know the logarithm of 2, we know also the logarithm of 20=2·10, or of 200=2·100, etc.

From the logarithm of 47, we have merely by changing the characteristic, the logarithms of 470, 4700, 47000, etc.

Thus we find by reference to the tables:

log.	2=0.3010300 ; also log.	47=1.6720979
"	20=1.3010300 ;	" 470=2.6720979
"	200=2.3010300 ;	" 4700=3.6720979
"	2000=3.3010300 ;	" 47000=4.6720979
"	200000=5.3010300 ;	" 47000000=7.6720979

266.

The logarithm of a quotient, is obtained by subtracting the logarithm of the divisor from the logarithm of the dividend. For this purpose, also, we consider all fractions as expressions of division. It therefore follows that when any number is to be divided by 10, 100, 1000, etc., we simply reduce the characteristic of its logarithm by as many units as there are zeros in the divisor and leave the mantissa unchanged.

Thus, for example:
$$\log. 4571 = 3.6600112$$
whence, $\log. \frac{4571}{10} = 2.6600112$
$$\log. \frac{4571}{100} = 1.6600112$$

This follows because:
$$\frac{4571}{100} = \frac{10^{3.6600112}}{10^2} = 10^{1.6600112}.$$

267.

From the foregoing paragraphs we can deduce at once the rule by which to find the logarithm of any whole number and decimal combined. Take first the characteristic belonging to the *whole* part of the number *only*. Then take from the tables the mantissa which corresponds to the number *including* the decimal *portion* as if there were no decimal point present. Thus:

$$\tfrac{4571}{10}=457.1\ ;\ \tfrac{4571}{100}=45.71, \text{ etc.}$$

Therefore: log. $457.1 = 2.6600112$
" $45.71 = \bar{1}.6600112$
" $4.571 = 0.6600112$

268.

When we have to find the logarithm of a whole number and common fraction, such for example as $36\tfrac{3}{4}$, we may proceed in two ways. We may convert the fraction into a decimal and proceed as in the preceding paragraph, or we may convert it into an improper fraction, and then subtract the logarithm of the denominator from the logarithm of the numerator.

Thus, since $36\tfrac{3}{4} = 36.75 = \tfrac{147}{4}$, we have:

$$\log.\ 36\tfrac{3}{4} = \log.\ 36.75 = 1.5652573$$

$$\text{or } \log.\ 36\tfrac{3}{4} = \begin{cases} \log.\ 147 = 2.1673173 \\ 4 = 0.6020600 \end{cases}$$

$$\overline{\log.\ \tfrac{147}{4} = 1.5652573}$$

269.

To take out the logarithms of proper fractions or of decimals from the table and determine their correct characteristics, we we must observe the following points: Since $10^0 = 1$, the exponent for any quantity less than 1 must be smaller than 0, and must be a proper fraction. Thus, for example:

$$10^{-1} = \tfrac{1}{10},\ 10^{-2} = \tfrac{1}{100}. \qquad \text{(See § 205)}.$$

The logarithms of all proper fractions must therefore be negative, and the smaller the value of the proper fraction the greater the absolute number of the corresponding negative logarithm.*

---

*When a fraction becomes infinitely small its negative logarithm must become infinitely large. A magnitude which becomes infinitely large, and can no longer be expressed by figures, is indicated by the sign ∞; and a quantity which has become infinitely small, so as to be indistinguishable from 0, is represented by $\tfrac{1}{\infty}$. These expressions also give rise to:

$$10^{-\infty} = \tfrac{1}{10^{\infty}} = 0 = \tfrac{1}{\infty}$$

Therefore we have: log. $0 = \bar{\ }\infty$ (Read "infinite negative",.

In order to write the negative logarithm of a true decimal (i. e., a decimal without any whole number attached) acording to § 266, we can unite the decimal with its proper denominator, and then subtract the logarithm of the denominator from the logarithm of the numerator.

Thus for example:

$$0.0564 = \tfrac{564}{10000}$$
$$\log. 564 = 2.7512791$$
$$\text{and } \log. 10000 = 4.0000000$$
$$\text{hence } \log. 0.0564 = -1.2487209$$
$$\text{for } \tfrac{564}{10000} = \frac{10^{2.7512791}}{10^4} = 10^{2.7512791-4} = 10^{-1.2487209}$$

270.

When a negative logarithm occurs, not as the final result of a calculation, but is to be added to or subtracted from other logarithms, or the number corresponding to it is to be determined, it is then found much more convenient to separate the negative logarithm into two parts, of which one part (the mantissa) shall consist of a *positive*, proper decimal fraction, and the other part a *negative* whole number, connected to the positive part as a negative characteristic.

A negative logarithm is easily brought into this form, by appending such a positive number, in connection with $+$ and $-$ signs, as shall convert the logarithm into one with a negative characteristic, and positive mantissa. This is done in the following manner:

Suppose, for example, that we have:

$$\log. 0.0564 = -1.2487209$$

we may add 2 and subtract 2, by which operation the actual value of the logarithm will not be altered, but only its form changed. Thus:

$$\log. 0.0564 = 2 - 1.2487209 - 2$$

Then collecting the first two members of the right-hand side together we have:

$$\log. 0.0564 = 0.7512791 - 2$$

giving us the required form, in which we have the positive mantissa with 0 for a characteristic, and also a negative whole number

(-2) for the negative characteristic. Instead of writing the negative characteristic after the mantissa, it is often written before it, in the usual place, but with the minus sign *over* the characteristic, in order to show that the whole number only is negative, and that the minus sign does not apply to the decimal portion, thus:

$$\log. \ 0.0564 = \bar{2}.7512791$$

271.

The denominator of a decimal fraction is equal to 1, with as many zeros appended as there are places in the decimal, thus:

$$0.564 = \tfrac{564}{1000}; \quad 0.0564 = \tfrac{564}{10000}, \text{ etc.}$$

We therefore see that the characteristic of the logarithm of the denominator will be greater than that of the numerator by as many units as there are zeros between the decimal point and the first significant figure of the decimal. Thus in the fraction $\tfrac{564}{1000}$ the characteristic of the numerator will be 2 and that of the denominator will be 3; while for $0.0564 = \tfrac{564}{10000}$, the characteristic of the denominator will be 4, while that of the numerator will still remain $= 2$.

From these considerations we obtain a simple rule to find the logarithm of a proper decimal fraction from the regular tables. We first find the logarithm of the number of which the decimal is composed, just as if the decimal point did not exist, and write it with a characteristic $= 0$, and then append a negative characteristic of as many units as there are zeros in the decimal plus 1. Thus:

$$\log. \quad 0.564 = 0.7512791 - 1, \text{ or } \bar{1}.7512791$$
$$\text{`` } \quad 0.0564 = 0.7512791 - 2, \text{ or } \bar{2}.7512791$$
$$\text{`` } \quad 0.00564 = 0.7512791 - 3, \text{ or } \bar{3}.7512791$$

This operation simply amounts to considering the decimal as a vulgar fraction with its corresponding denominator of 100, 1000, etc., and then taking the logarithm of the numerator and subtracting that of the denominator, except that the subtraction is *indicated* instead of being performed.

272.

In order to find the logarithm of a proper vulgar fraction, the fraction may be converted into a decimal and the logarithm found at once by the above rule. Thus we have:

$$\log. \tfrac{7}{16} = \log. 0.4375 = 0.6409781 - 1$$
$$\text{or,} = \bar{1}.6409781$$

Frequently it is more convenient to subtract the logarithm of the denominator from the logarithm of the numerator, as the following example shows:

$$\log. 7 = 0.8450980$$
$$\text{`` } 16 = 1.2041200$$
$$\log. \tfrac{7}{16} = \bar{1}.6409780$$

Here we have a positive mantissa, being the difference between the two mantissas, while the characteristic is $0-1=-1$, as before.

The practice of placing the characteristic in the usual place, with the negative sign over it, is much the more convenient in practice, and will be used altogether hereafter.

Again we have:

$$\log. \tfrac{3}{7} = \begin{cases} \log. 3 = 0.4771213 \\ \log. 7 = 0.8450980 \end{cases}$$
$$\bar{1}.6320233$$

Here the lower decimal is greater than the upper, and hence 1 was borrowed from the unit's place, which gives -1 in the result.

Again:

$$\log. \tfrac{11}{4771} = \begin{cases} \log. 11 = 1.0413927 \\ \log. 4771 = 3.6786094 \end{cases}$$
$$\bar{3}.3627833$$

Here the lower decimal is the greater, and 1 is borrowed from the unit's place of the subtrahend, leaving 0, and then $0-3=-3$ for the negative characteristic.

Whenever circumstances may require it, we can modify any logarithm by adding, *at the same time*, equal positive and negative characteristics. Thus:

$$\log. \tfrac{3}{7} = \quad \bar{1}.6320233$$
$$-6+6$$
$$= -6+5.6320233$$

or   log. $\frac{\rho}{q} =$ 1.6320233
                    —4+4
                    —4+3.6320233   etc.

The use of this will be seen hereafter.

273.

Bearing in mind the rules given in the preceding paragraphs we will now proceed to explain more fully the method of taking out from the tables the logarithms of any required numbers:

If we open any table of seven figure logarithms, such as Schrön's, we see at the left of the page the column headed Num. (Numbers). Then following toward the right 10 columns, headed successively 0, 1, 2, 3, 4, 5, 6, 7, 8, 9, and at the right a column headed P. P. (Proportional Parts).

In some tables the column of numbers is simply headed N, and the column of proportional parts is headed, "Differences."

(1) Turning now, for example, to the number 1267 in the column "Num." we find opposite 1267 in the column 0, the logarithm .1027766, or with the characteristic, 3.1027766. We do not find the first three figures of the decimal repeated for every number, but only printed in where they change, hence when there is a space for the first three figures those which are next above are to be taken. Likewise the succeeding columns contain only four figures, and these are in each case to be preceded by the corresponding three figures printed in the 0 column. This saves the constant repetition of the first three figures, and economizes space very much.

If a number contains only four figures its logarithm is found at once opposite to the number in the 0 column, but if the number consists of five figures the fifth figure must be found in the column headings at the top of the page, and in the corresponding column and on the same horizontal line as the first four figures will be found the proper logarithm. Thus we have :

log. 1267=3.102 7766   log. 12675=4.102 9480
log. 12670=4.102 7766   log. 12676=4.102 9822
log. 12671=4.102 8109   log. 12677=4.103*0165
log. 12672=4.102 8452   log. 12678=4.103*0507
log. 12673=4.102 8974   log. 12679=4.103*0850
log. 12674=4.102 9137

Examining these we see that the logarithms of 1267 and 12670 have the same mantissa, the characteristic being simply increased by 1. If the fifth figure, however, is greater than 0 we take out the last four decimals from the column headed by this figure, the first three figures of the mantissa remaining unchanged. In some cases, as log. 12677, the third figure also is changed, and this is indicated by an asterisk (*) in front of the fourth figure. When this * appears the first three figures of the mantissa are to be taken from the next line *below* in the zero column. Thus when we look for the log. 12677 we have 102 in the 0 column, and *0165 in column 7, and the asterisk means that instead of 102 we must take 103 for the first three figures. In some tables a dot is used instead of an asterisk, and in others a dash (—), but the meaning is the same in all.

We therefore have the following rule to take out the logarithms of five figure numbers.

First write the proper characteristic, then find the first four figures of the given number in the column "Num," and opposite these in the 0 column find the first three figures of the mantissa. Then find the last four figures of the mantissa in the column headed by the fifth figure of the given number, on the same horizontal line as the first four figures were found. If there is also an asterisk here found, the third figure of the mantissa is to be increased by 1.

If the given five figure number is multiplied or divided by 10, 100, 1000, etc., the mantissa remains unchanged and the characteristic alone is altered. Thus:

$$\log. 22035 = 4.3431131$$
$$\log. 2.2035 = 0.3431131$$
$$\log. 33829 = 4.5292892$$
$$\log. 338.87 = 2.5300331$$
$$\log. 338.29 = 2.5292892$$
$$\log. 78.164 = 1.8930068$$
$$\log. 781640 = 5.8930068$$
$$\log. 0.049097 = \bar{2}.6910550$$
$$\log. 1.1011 = 0.0418268$$

274.

If we take out the logarithms of several successive five-figure numbers, and subtract them from each other, we find the differences appear only in the last three figures, and that for some little space these differences are all equal to each other.

Thus for example:

$$\log. 23740 = 4.3754807 \quad \text{Difference}$$
$$\log. 23741 = 4.4754990 \quad 183$$
$$\log. 23742 = 4.3755173 \quad 183$$
$$\log. 23743 = 4.3755356 \quad 183$$
$$\log. 33744 = 4.3755539 \quad 183$$
$$\log. 23745 = 4.3755722 \quad 183$$

We thus see that if we increase the number 23740 by 1, we increase the last 3 decimals of its logarithm by 183, and if the number 23740 increases 2, 3, or 4 units, we must add 2, 3, or 4 times 183 to the last figures of its logarithm.

Since this proportion exists for the increase of the number by units, it must also hold good for fractional parts of units, such as $\frac{1}{10}$, $\frac{2}{10}$, $\frac{1}{100}$, $\frac{2}{100}$, etc. If, therefore, the number 23743, for example, is increased by $\frac{1}{10}$, $\frac{2}{10}$, $\frac{1}{100}$, etc., then also will the last decimals of the logarithm be increased by $\frac{1}{10}$, $\frac{2}{10}$, $\frac{1}{100}$, etc., of the difference, namely by $\frac{1}{10} \cdot 183$, or $\frac{2}{10} \cdot 183$, etc.

If therefore we know the difference between the logarithm of 23743 and the next logarithm, we can find the logarithms of $23743\frac{8}{10} = 23743.8$; or of 23743.85; or of 23743.859 etc., or of a number 10, 100, or 1000 times as great, such as $237438 = 10 \cdot 23743.8$; or $2374385 = 100 \cdot 23743.85$.

Suppose for example that we have:

$$\log. 23743 = 4.3755356$$

If now we add to the last decimals 0.8 times the difference 183, i. e.,

$$0.8 \cdot 183 = 146.4, \text{ say } 146$$

We have: $\qquad$ 4.3755356
$\qquad\qquad\qquad\qquad$ 146

$$\log. 23743.8 = 4.3755502$$
$$\text{and} \quad \log. 237438 = 5.3755502$$

If we add 0.85 times the difference 183, we have:

$$0.85 \cdot 183 = 155.55, \text{ say } 156$$

and $\qquad\qquad$ 4.3755356
$\qquad\qquad\qquad\qquad$ 156

$$\log. 23743.85 = 4.3755512$$
$$\text{and} \quad \log. 2374385 = 6.3755512 \quad \text{etc.}$$

In the best seven figure tables, such as Schrön's, this calculation is rendered unnecessary, for in the column headed P. P. (at the extreme right) are given the products of the differences by the digits from 1 to 9, so that the amount to be added for any required number can be taken at once.

Thus we find for
$$\log. 23743859 = 7.3755513$$
in the following manner:
$$\text{The } \log. 23743 = 4.3755356$$
then for the remaining figures we have:

$$\tfrac{859}{1000} \cdot 183 = (\tfrac{8}{10} + \tfrac{5}{100} + \tfrac{9}{1000}) 183 =$$
$$= \tfrac{8}{10} \cdot 183 + \tfrac{1}{10} \cdot \tfrac{5}{10} \cdot 183 + \tfrac{1}{100} \cdot \tfrac{9}{10} \cdot 183$$

In the column of Proportional Parts we now find:

$\tfrac{8}{10} \cdot 183 = 146.4$

$\tfrac{5}{10} \cdot 183 = 91.5$ hence $\tfrac{1}{10} \cdot 91.5 = 9.15$

$\tfrac{9}{10} \cdot 183 = 164.7$ hence $\tfrac{1}{100} \cdot 164.7 = 1.647$

and we have: log. 23743... = 4.3755356
                    8   146.4
                    5.   9.15
                    9   1.645

    log. 23743.859 = 4.3755513
and log. 23743859 = 7.3755513
or log. 237.43859 = 2.3755513

In the same way we find the logarithm of 1275.8073. Thus:
    log. 1275.8... = 3.1057826
           0..   0
           7.   23.8
           3   1.02
    log. 1275.8073 = 3.1057851

275.

When we desire to indicate the number corresponding to a given logarithm, we unite num. log: (i. e. *numerus logarithmi.*)

Thus since log. 2 = 0.3010300

We have inversely:
$$\text{num. log. } 0.3010300 = 2$$
which is read "the *number* for the logarithm 0.3010300 is 2."

The method of finding the number from the table, when the logarithm is given, is easily deduced from the preceeding rules.

(1) First look for the first three decimals of the mantissa in the column headed 0. Then look across horizontally on the same line or on the next lines immediately below, for the last four decimals, in the column headed from 0 to 9. If these last four are not found here, look one line higher, in which case they must be preceded by an asterisk. If they are not found *exactly*, take the nearest value. We then have in column Num. the first four figures of the required number. We then examine the characteristic of the given logarithm and give our number as many *whole* places as there are units in the characteristic plus 1, and thus place the decimal point.

If the logarithm has a negative characteristic, the number must all be a decimal, and there must be as many zeros placed at the left of the number as there are units in the characteristic less 1, and then the decimal point is placed.

This gives us the number correct to five figures, and the method can readily be understood by the student if he will take his volume of tables and find the logarithms corresponding to the following numbers and vice versa.

log. 22035=4.3431131
num. log. 4.3431131=22035

log. 666.42=2.8237480
num. log. 3.8237480=6664.2

log. 8.7707=0.9430343
num. log. 6.9430343=8770700

log. 0.92904=1.9680344
num. log. 3.9680344=0.0092904

log. 0.051001=2.7075787
num. log. 0.0010411=1.0024

(2) In order to find the additional figures when the exact value of the base decimals of the mantissa cannot be found in the table, we proceed as follows :

First find the first three decimals of the mantissa in column 0, as before. Then find in the other columns the *next lower* value to the four last figures, and take the head of the column for the fifth figure, as before.

Then subtract these next lower figures from the last four figures of the given logarithm and look for the nearest value in the corresponding portion of the column of proportional parts. The figure in the P. P. column which corresponds to this difference most closely will be the sixth figure of the number, always taking the next lower value. Then subtract this lower value from the tabular proportion thus taken, and the proportion corresponding to this difference will give the seventh figure. With seven place tables this is as far as the result can be accurately carried, and if the characteristic demands another figure its place must be filled by a zero.

An example will make the procedure clear.

Required : Num. log. 7.3255512.

We find in the table the next smaller logarithm is 7.3755356, hence :

$$\begin{array}{ll} \text{given log.} = 7.3755512 & \\ \text{next smaller log.} = 7.3755356 & \\ \text{and its number} = & 23743000 \\ \hline \text{Difference} & 156 \end{array}$$

In the column P. P. the next smaller value is 146.4, and the corresponding figure is 8. Then 156—146.4=9.6 which multiplied by 10=96, and the next smaller value is 91.5, for which the number is 5.

Hence : Num. log. 7.3655512=23743850
Also : Num. log. 4.3755512=23743 85

This rule follows evidently from what has been said before. The reasoning is as follows :

The difference between the given logarithm and that for a number smaller by 1 is 183. But the difference between the given logarithm and the next smaller logarithm is 156. Therefore we have the proposition :

$$\text{As } 183:1 \text{ so is } 156:x$$
$$x = \tfrac{156}{183} = 0.85$$

276.

The following examples will answer for practice, and larger numbers will rarely occur in actual examples. The 8 and

9 figure numbers will vary slightly from exactness beyond the seventh decimal, as greater accuracy can hardly be obtained with seven place tables.

The methods of using 5 and 6 figure tables of logarithms will readily be understood from what has been said above, and specific directions for the various editions will be found with them.

EXAMPLES.

(1) log. 370978 = 5.5693482
(2) num. log. 3.8911459 = 7782.98
(3) log. 8689836 = 6.9390116
(4) num. log. 6.9720151 = 9375946
(5) log. 200.36084 = 2.3018128
(6) num. log. 0.0692746 = 1.172937
(7) log. 0.07787009 = 2.8913707
(8) num. log. 3.0911392 = 0.0012335
(9) log. 4.501000895 = 0.6533091
(10) num. log. 0.0901392 = 1.230664

The Applications of Logarithms.

277.

In the following pages will be given examples and rules for the applications of logarithms, and among these, examples will be given of problems which without logarithms could only be solved with greatest difficulty, or in some cases could not be solved at all.

278.

To obtain the product of two numbers we add the logarithms of the factors; the sum is then the logarithm of the product, which can be obtained from the table as explained in § 258, 1. Stated as a general formula with letters for symbols we have:

$$x = abc$$

in which a, b, and c are factors and x is their product:

$$\log. x = \log. a + \log. b + \log. c$$

EXAMPLE. Find x from the following equation:

$$x = 823 \cdot 1305 \cdot \tfrac{3}{7} \cdot (2.40067)(0.0067925)$$

We have:

$$\log. 823 = 2.9153998$$
$$\log. 1305 = 3.1156105$$
$$\log. \tfrac{3}{7} = \log. 0.4285714 = \overline{1}.6320218_{10}^{4}$$
$$\log. 2.40067 = 0.3803198_{127}$$
$$\log. 0.0067925 = \overline{3}.8320296$$
$$\log. x = 3.8753956$$
$$x = 7505.776$$

279.

In order to divide one number by another, subtract the logarithm of the divisor from the logarithm of the dividend; the remainder will be the logarithm of the quotient.

If $x = \dfrac{a}{b}$ in which a is the dividend, b the divisor and x the quotient then we have:

$$\log. x = \log. a - \log. b.$$

EXAMPLE. Find x from the following equation:

$$x = \frac{25.0035}{7123.0409}$$

$$\log. 25.0035 = 1.3980008$$
$$\log. 7123.0409 = 3.8526653$$
$$\log. x = \overline{3}.5453355$$
$$x = 0.003510229$$

In the subtraction we had to borrow 1, from the units place, and then 0—3, gave $\bar{3}$ for the characteristic.

If either the dividend or divisor, or both, consist of factors, we first find the logarithms of the factors, and add them, and then perform the subtraction. Thus:

$$x = \frac{abc}{de}$$

$$\log. x = \log. (abc) - \log. (de)$$
or $\log. x = \log. a + \log. b + \log. c - (\log. d + \log. e)$
or $\log. x = \log. a + \log. b + \log. c - \log. d - \log. e$

EXAMPLE. Find x from the following equation:

$$x = \frac{0.035689 \cdot 6.083769}{34.595 \cdot 0.0050602}$$

We have for the numerator:

$$\log. 0.035689 = \bar{2}.5525344$$
$$\log. 6.083769 = 0.7841727$$
$$\overline{1.3367071}$$

for the denominator:

$$\log. 34.595 = 1.5390133$$
$$\log. 0.0050602 = \bar{3}.7041677$$
$$\overline{1.2431810}$$

$$\text{whence:} \quad 1.3367071$$
$$1.2431810$$
$$\log. x = 0.0935261$$
$$x = 1.240298$$

280.

In order to raise a number to a given power, the logarithm of the number is multiplied by the exponent of the power; the product will be the logarithm of the power. (See § 258, 3).

Thus:

log. a^4 = log. $aaaa$ = log. a + log. a + log. a + log. a whence log. a^4 = 4 log. a.

In general:
$$\log. a^n = n \log. a$$
$$\text{or} \quad \log. a^x = x \log. a$$

EXAMPLE 1. Find x from the following equation:
$$x = (1.3504)^{22}$$
We have:
$$\log. 1.3504 = 0.1304624$$
$$22$$
$$\overline{}$$
$$2609248$$
$$2609248$$
$$\overline{}$$
$$\log. x = 2.8701728$$
$$x = 741.6052$$

EXAMPLE 2. Find x from the following:
$$x = (\tfrac{200}{331})^{10}$$
$$\log. 200 = 2.3010300$$
$$\log. 331 = 2.5198280 \quad (\S\ 272)$$
$$\overline{1.7812020}$$
$$10$$
$$\overline{}$$
$$\log. x = \overline{3}.8120200$$
$$x = 0.006486643$$

Here we have the positive mantissa
$$0.7812020 \times 10 = 7.8120200$$
and the negative characteristic $\overline{1} \times 10 = \overline{10}$.

$$\text{hence log. } x = \overline{3}.8120200$$

281.

In order to extract any given root we divide the logarithm of the number by the exponent of the root; the quotient will be the logarithm of the root.

Thus (§258,4):
$$\log. \sqrt[5]{a} = \log. a^{\frac{1}{5}} = \tfrac{1}{5}\log. a$$
and in general:
$$\log. \sqrt[n]{a} = \frac{1}{n} \log. a$$

EXAMPLE.
$$x = \sqrt[7]{2}$$
$$\log. 2 = 0.3010300$$
$$7)0.3010300$$
$$\log. x = 0.0430043$$
$$x = 1.104089$$

If we have to extract a root of a proper fraction, i. e., one of which the logarithm has a negative characteristic, we must add as many units to the negative characteristic as will make it positive and will also be exactly divisible by the exponent; also, of course, subtracting the same number of units, in order that the value of the logarithm be not altered. We then divide both quantities by the given exponent, as follows:

EXAMPLE 2. Required:
$$x = \sqrt[5]{0.0375}$$
$$\log. 0.0375 = \overline{2}.5740313$$
adding —5+5.
$$5) \overline{-5 + 3.5740313}$$
$$-1 + 0.7148063$$
$$\log. x = \overline{1}.7148063$$
$$x = 0.5185687$$

282.

If we have to multiply two or more factors together, which factors are also powers, we take the logarithms of the roots, and multiply them first by their respective exponents, and then add the results, to effect the multiplication, as will appear by the following examples :

(1) $$x = a^3 b^2 = aaa.bb$$
log. $x =$ log. $a +$ log. $a +$ log. $a +$ log. $b +$ log. b
or log. $x = 3$ log. $a + 2$ log. b

(2) $$x = a^m b^n c^n$$
log. $x = m$ log. $a + n$ log. $b + n$ log. c

(3) $$x = a^{\frac{m}{n}} \sqrt[n]{b^m} a^{\frac{m}{n}} b^n$$
log. $x = \dfrac{m}{n}$ log. $a + \dfrac{n}{m}$ log. b

283.

It is evident that we can only obtain a product by means of logarithms when the factors are monomials. When the factors consist of several members, each factor must be considered as a single member, or if the various members are numbers they must be collected into single members before the logarithmic calculations are made. Thus, for example :

(1) $$x = a^m (a+b)^n$$
log. $x = m$ log. $a + n$ log. $(a+b)$

and $a+b$ must be determined before its logarithm can be taken.

(2) $$x = \dfrac{(a+b^m)^n (a+b)}{(a-1)^n}$$
log. $x = n$ log. $(a+b^m) +$ log. $(a+b) - n$ log. $(a-1)$

(3) $$x = \sqrt[15]{(1\tfrac{15}{32} - \sqrt[5]{11\tfrac{3}{44}})}$$

log. $3 = 0.4771213$
log. $1144 = 3\ 0584260$

$\phantom{\text{log. }1144=}3\ 4186953$
$\phantom{\text{log. }1144=3\ }-5+5.$
$\phantom{\text{log. }}5)\overline{-5+2.4186953}$

num. log. $-1 + 0.4837391 = 0.3046064$

which subtracted from $1\tfrac{15}{32} = 0.46875$

gives $1\tfrac{15}{32} - \sqrt[5]{11\tfrac{3}{44}} = 0\ 1641436$

$$\text{log. } 0.1641436 = \overline{1}.2152240$$
$$-15 + 15$$
$$15) \overline{} - 15 + 14.2152240$$
$$\overline{1} + 0.9476816$$
$$\text{log. } x = \overline{1}.9476816$$
$$x = 0.886506$$

284.

Since we are able by means of logarithms, readily to extract any root whatever, it is also clear that we may use them to solve all pure equations of the higher degrees. Suppose, for example, that it is required to find the value of x, from the following equation, (§ 219):

$$\tfrac{3}{4}x^{11} + 0.501 = 2x^{11} + 6.05$$

This is a pure equation of the 11th degree.
Collecting the values of x^{11}, we have:

$$\tfrac{3}{4}x^{11} - 2x^{11} = -0.501 + 6.05$$
$$-\tfrac{5}{4}x^{11} = 5.549$$
$$x = \sqrt[11]{-4.16175} \qquad (\S\ 216,3)$$
$$\text{log. } 4.16175 = 0.6192760\ (n)$$
$$\text{log. } x = \frac{0.6192760\ (n)}{11} = 0.0562978$$
$$x = -1.138407$$

The (n) appended to the logarithm means that the number to which it belongs has the minus sign. Since the logarithms of whole numbers are positive, and those of proper fractions are negative, it is evident that we cannot find in the tables, logarithms for negative numbers. In order therefore to work with logarithms for negative numbers, we consider the numbers as positive, and append the symbol (n) to the logarithms, and give the negative sign to the result.

285.

By the use of logarithms it is often easy to determine the unknown quantity in an equation, when the unknown appears as an exponent. Suppose, for example, it is required to find what power of 2 will equal 64, we have the following equation:

$$2^x = 64$$

Taking the logarithms of both sides, we have :
$$x \log 2 = \log 64$$
$$x = \frac{\log 64}{\log 2} = \frac{1.8061800}{0.3010300} = 6$$

The student must be careful not to confuse $\frac{\log 64}{\log 2}$ with log. $\frac{64}{2}$ = log. 64 — log. 2. In the latter case we subtract one logarithm from the other, but in the former case we divide the one by the other. In this latter case if desired we can consider the logarithms as ordinary numbers, taking again the logarithms of each of them from the table and subtracting the logarithm of the divisor from the logarithm of the dividend to obtain the logarithm of the quotient, and thus save the labor of the long division.

In general, when :
$$a^x = b$$
$$x \log. a = \log. b$$
$$x = \frac{\log. b}{\log. a}$$

Again if :
$$a^{mx} b^{n-\frac{x}{2}} = c^{k-x} c$$

We have :
$$mx \log. a + (n - \tfrac{x}{2}) \log. b = (k - x) \log. c + \log. d$$
$$xm \log. a + n \log. b - \tfrac{x}{2} \log. b = k \log. c - x \log. c + \log. d$$
$$xm \log. a - \tfrac{x}{2} \log. b + x \log. c = k \log. c + \log. d - n \log. b$$
$$x = \frac{k \log. c + \log. d - n \log. b}{m \log. a + \log. c - \tfrac{1}{2} \log. b}$$

286.

EXAMPLE 1. Required x, from the equation :
$$a \cdot c^{mx} - b \cdot c^{\frac{mx}{2}} = d$$

Solution. In order to simplify the work let us temporarily represent the $c^{\frac{mx}{2}} = z$, whence $c^{mx} = z^2$, and we have (§ 231)
$$az^2 - bz = d$$
$$z = \frac{b \pm \sqrt{b^2 + 4ad}}{2a}$$

replacing z by its value $c^{\frac{mx}{2}}$, we have:

$$c^{\frac{mx}{2}} = \frac{b \pm \sqrt{b^2+4ad}}{2a}$$

$$\frac{mx}{2}\log. c = \log. \left(\frac{b \pm \sqrt{b^2+4ad}}{2a}\right)$$

$$x = \frac{2}{m \log. c} \cdot \log. \left(\frac{b \pm \sqrt{b^2+4ad}}{2a}\right)$$

EXAMPLE 2. Find x, from the equation:

$$a^x + \frac{1}{a^x} - b = 0$$

Solution. Multiplying by a^x, and then putting $a^x = z$, we readily find, according to the method given in § 231:

$$x = \frac{\log. \left[\frac{1}{2}(b \pm \sqrt{b^2-4})\right]}{\log. a}$$

287.

EXAMPLE 3. Interpolate four terms between 2 and 8, so that all six terms shall form a geometrical progression.

Solution. Given $a=2$, $l=8$, $n=6$, and e, required.

We have from § 253:

$$l = ae^{n-1}$$

whence:

$$e = \sqrt[n-1]{\frac{l}{a}}$$

$$\log. e = \frac{\log. l - \log. a}{n-1}$$

$$\log. e = \frac{\log 8 - \log. 2}{n-1}$$

$$e = 1.319508$$

The series will therefore be:

$$2\ ;\ 2\cdot(1.319)\ ;\ 2\cdot(1.319)^2\ ;\ 2\cdot(1.319)^3\ ;\ 2(1.319)^4\ ;\ 8.$$

288.

EXAMPLE 4. It is related that Sissa, the inventor of chess, was requested by the Hindu Rajah Scheran, to demand a reward for his ingenious invention. Sissa modestly requested that he should have the sum of the number of grains of rice that could be counted by the squares of the chess board, if counted in the following manner. On the first square was to be placed 1 grain, on the second square two grains, on the third square 4 grains, on the fourth square 8 grains, &c., increasing in this progression: $1+2+4+8+16\ldots+2^{63}$; that is doubling on each square the amount on the preceding square until the whole 64 squares had been counted. The rajah was at first offended at a demand upon him so small as to be beneath his dignity, but ordered the gift to be made, when he was surprised to find it far beyond the power of himself or any other monarch. How much would the total be?

Solution. Given $a=1$, $e=2$, $l=2^{63}$, and s required.

We have from § 254.:

$$s = \frac{le-a}{e-1}$$

$$s = \frac{2^{63} \cdot 2 - 1}{2-1} = 2^{64} - 1$$

It will be close enough to assume:

$$s = 2^{64}$$

whence log. $s = 64$ log. 2

$s = 18446750000000000000000$

More exactly, the sum would be

$s = 18446744073709551615$

The last figures cannot be found by means of logarithms, as they exceed the limits of any table which has ever been calculated. It is estimated that if the whole surface of the earth was under cultivation it would require the harvests of seventy years to equal this amount.

289.

EXAMPLE 5. A wine butt contains 100 gallons of wine, which we will call a. From this 1 gallon $=b$, is drawn, and an equal amount of water poured in. When the water and wine are thoroughly mixed the amount b is again drawn and replaced with water. After this operation has been repeated 20 times, how much wine remains?

Solution. The problem may be stated in the following manner: If 1 gallon is taken from 100 gallons and replaced by water there must remain 99 gallons of wine, and the mixture will consist of $\frac{1}{100}\cdot 99$ of wine and $\frac{1}{100}$ of water. When the second draught is made, there will remain $99-\frac{1}{100}\cdot 99$ of wine and $1-\frac{1}{100}\cdot 1$ of water. After the third draught there will be $(99-\frac{1}{100}\cdot 99)$ of pure wine, etc., etc.

In general, therefore, if $a =$ the original amount of wine, and b the amount of each draught, we have taken the proportion $\frac{b}{a}$ of the original amount the first time and left $a - \frac{b}{a} = a - b$ in the the vessel. After putting in the water, we take out the second time $\frac{b}{a}(a-b)$ of wine and there will remain:

$$(a-b) - \frac{b}{a}(a-b) = \frac{a(a-b)-b(a-b)}{a} = \frac{(b-b)^2}{a}$$

After the third time we have left:

$$\frac{(a-b)^2}{a} - \frac{b}{a}\cdot\frac{(a-b)^2}{a} = \frac{(a-b)^3}{a^2}$$

After the fourth time $\frac{(a-b)^4}{a^3}$, etc.

If x be the amount of wine remaining after the nth draught we have in general:

$$x = \frac{(a-b)^n}{a^{n-1}}$$

$$\log. x = n \log. (a-b) - (n-1) \log. a$$

If, as in the example: $a = 100$, $b = 1$, $n = 20$, we have

$$\log. x = 20 \log. 99 - 19 \log. 100$$

$$x = 81.79072$$

LOGARITHMS.

150. Def.—The **logarithm** of a number to a given base is the index of the power to which the base must be raised to obtain the number.

Thus, we may obtain the numbers 1, 10, 100, 1,000, 10,000, &c., by raising the base 10 to the powers 0, 1, 2, 3, 4, &c., respectively; and hence, by the above definition, we have—

$\text{Log}_{10} 1 = 0$, $\log_{10} 10 = 1$, $\log_{10} 100 = 2$, $\log_{10} 1,000 = 3$, &c.,

the suffix 10 being added to the word *log* to indicate that the base is 10. It is usual, however, in common logarithms to omit this suffix; and hence, when there is no base expressed, the student will understand 10.

Again, the numbers 1, 2, 4, 8, 16, &c., may be obtained by finding the values of 2^0, 2^1, 2^2, 2^3, 2^4, &c., respectively, and hence we have by definition—

$\text{Log}_2 1 = 0$, $\log_2 2 = 1$, $\log_2 4 = 2$, $\log_2 8 = 3$, &c.

So also we find $\log_4 16 = 2$, $\log_5 125 = 3$, $\log_3 81 = 4$, &c.

Ex. Find $\log_4 256$, $\log_{36} 216$, and the logarithm of 9 to base $\sqrt{3}$.

$\text{Log}_4 256 = \log_4 4^4 = 4$, by definition.

$\text{Log}_{36} 216 = \log_{36} 6^3 = \log_{36} (6^2)^{\frac{3}{2}} = \log_{36} 36^{\frac{3}{2}} = \frac{3}{2}$, by definition.

$\text{Log}_{\sqrt{3}} 9 = \log_{\sqrt{3}} 3^2 = \log_{\sqrt{3}} (\sqrt{3})^4 = 4$, by definition.

Characteristics of Ordinary Logarithms.

151. Def.—The **characteristic** of a logarithm is the integral part of the logarithm, and the fractional part (generally expressed as a decimal) is called the **mantissa.**

ALGEBRA.

1. *Numbers containing integer digits.*

Every number containing n digits in its integral part must lie between 10^{n-1} and 10^n.

Thus, 6 lies between 10^0 and 10^1, 29 lies between 10^1 and 10^2, 839 lies between 10^2 and 10^3, &c.

Hence the ordinary logarithms of all numbers having n integer digits lies between $(n - 1)$ and n.

The integral portion or characteristic of the logarithm of a number having n integer digits is therefore $(n - 1)$.

Hence we have the following rule :—

RULE 1.—*The characteristic of the logarithm of a number having integer digits is* ONE *less than the number of integer digits.*

Thus, the characteristics of the logarithms of 32, 713·54, 8·7168, 56452, 73607·9 are respectively 1, 2, 0, 4, 4.

2. *Numbers less than unity expressed as decimals.*

All such numbers having n zeros immediately after the decimal point lie between $\dfrac{1}{10^n}$ and $\dfrac{1}{10^{n+1}}$, or between 10^{-n} and $10^{-(n+1)}$.

Thus, ·3 lies between 1 and ·1, or between 1 and $\dfrac{1}{10}$, or 10^0 and 10^{-1};

·027 lies between ·1 and ·01, or between $\dfrac{1}{10}$ and $\dfrac{1}{10^2}$, or 10^{-1} and 10^{-2};

·000354 lies between ·001 and ·0001, or between $\dfrac{1}{10^3}$ and $\dfrac{1}{10^4}$, or 10^{-3} and 10^{-4}, and so on.

Hence, by Def., Art. 150, the logarithm of any number having n zeros immediately after the decimal point lies between $- n$ and $- (n + 1)$. Hence, the logarithm is *negative*, and the integral part of this negative quantity is n. It is however usual to write all the mantissæ of logarithms as *positive* quantities, and the *negative integral* part of the logarithm will be the next higher negative integer, viz., $- (n + 1)$.

We have therefore the following rule :—

RULE 2.—*The characteristic of the logarithm of a number less than unity, and expressed as a decimal, is the negative integer next greater than the number of zeros immediately after the decimal point.*

Thus, the characteristics of the logarithms of $\cdot 3$, $\cdot 0076$, $\cdot 02535$, $\cdot 7687$, are respectively -1, -3, -2, -1.

152. *The logarithm of the* PRODUCT *of two numbers is the* SUM *of the logarithms of the numbers.*

Let m and n be the numbers, and let a be the base. Since m and n must be each some power of a, integral or fractional, positive or negative, assume—

$$\left. \begin{array}{l} m = a^x, \\ n = a^y. \end{array} \right\} \text{ Then, by definition of a logarithm,}$$
$$x = \log_a m, \text{ and } y = \log_a n.$$

Now we have $mn = a^x \cdot a^y = a^{x+y}$, and hence, by definition,

$$\log_a mn = x + y; \text{ we therefore have—}$$

$$\log_a (mn) = \log_a m + \log_a n. \quad Q.E.D.$$

COR. This proposition may be extended to any number of factors.

Thus, $\log_a (mnpq) = \log_a m + \log_a n + \log_a p + \log_a q$.

153. *The logarithm of the* QUOTIENT *of two numbers is found by* SUBTRACTING *the logarithm of the denominator from the logarithm of the numerator.*

Assuming, as in the last Art., we have—

$$x = \log_a m, \; y = \log_a n.$$

Also, $\dfrac{m}{n} = \dfrac{a^x}{a^y} = a^{x-y}$, and hence, by definition,

$$\log_a \frac{m}{n} = x - y; \text{ we therefore have—}$$

$$\log_a \frac{m}{n} = \log_a m - \log_a n. \quad Q.E.D.$$

154. *The logarithm of the* POWER *of a number is found by* MULTIPLYING *the logarithm of the number by the* INDEX *of the power.*

ALGEBRA.

Let it be required to find $\log_a N^p$.

Assume $N = a^x$, and therefore $x = \log_a N$.

We have $N^p = (a^x)^p = a^{px}$, and hence, by definition,
$$\log_a N^p = px = p \log_a N. \quad Q.E.D.$$

Ex. 1.—Given $\log 2 = \cdot3010300$, and $\log 3 = \cdot4771213$, find the logarithms of 18, 15, ·125, 6·75.

Log $18 = \log (2 \times 3^2) = \log 2 + 2 \log 3 = \cdot3010300 + 2(\cdot4771213) = 1\cdot2552726$.

Log $15 = \log (3 \times \frac{10}{2}) = \log 3 + \log 10 - \log 2 = \cdot4771213 + 1 - \cdot3010300 = 1\cdot1760913$.

Log $\cdot125 = \log\left(\frac{1}{2^3}\right) = \log 1 - 3 \log 2 = 0 - 3 \times \cdot3010300$
$= -\cdot9030900 = -1 + (1 - \cdot9030900) = -1 + \cdot0969100$,
or, as usually written, $= \overline{1}\cdot0969100$.

Log $6\cdot75 = \log\frac{27}{4} = \log\frac{3^3}{2^2} = 3\log 3 - 2\log 2 = 3(\cdot4771213)$
$- 2(\cdot3010300) = \cdot8293039$.

Ex. 2. Find the logarithm of $\dfrac{(2\cdot4)^{\frac{1}{2}} \times (\cdot375)^4}{(2\cdot43)^5 \times (\cdot032)^{-\frac{1}{3}}}$, having given $\log 2$ and $\log 3$.

We have—

Log $N = \frac{1}{2} \log 2\cdot4 + 4 \log \cdot375 - 5 \log 2\cdot43$
$\qquad - (-\frac{1}{3}) \log \cdot032$.

$= \frac{1}{2} \log \frac{2^3 \times 3}{10} + 4 \log \frac{3}{2^3} - 5 \log \frac{3^5}{10^2} + \frac{1}{3} \log \frac{2^5}{10^3}$

$= \frac{1}{2} (3 \log 2 + \log 3 - \log 10) + 4 (\log 3 - 3 \log 2)$
$\quad - 5 (5 \log 3 - 2 \log 10) + \frac{1}{3} (5 \log 2 - 3 \log 10)$

$= (\frac{3}{2} - 12 + \frac{5}{3}) \log 2 + (\frac{1}{2} + 4 - 25) \log 3$
$\qquad + (-\frac{1}{2} + 10 - 1) \log 10$

$= -\frac{53}{6} \times \cdot3010300 - \frac{41}{2} \times \cdot4771213 + \frac{17}{2} \times 1$

$= -2\cdot6590983 - 9\cdot7809867 + 8\cdot5 = -3\cdot9400850$

$= \overline{4} + (4 - 3\cdot9400850) = \overline{4}\cdot0599150$.

The Use of Tables.

155. Tables have been formed of the logarithms of all numbers from 1 to 100,000, and we shall now show how they are practically used. We shall not enter here upon the method of forming the tables themselves.

The following is a specimen of the way in which the logarithms of numbers are usually tabulated:—

No.	0	1	2	3	4	5	6	7	8	9	D.
7990	9025468	5522	5577	5631	5685	5740	5794	5848	5903	5957	
91	6011	6066	6120	6174	6229	6283	6337	6392	6446	6500	
92	6555	6609	6663	6718	6772	6826	6881	6935	6989	7044	
93	7098	7152	7207	7261	7315	7370	7424	7478	7533	7587	
94	7641	7696	7750	7804	7859	7913	7967	8022	8076	8130	54
95	8185	8239	8293	8348	8402	8456	8511	8565	8619	8674	
96	8728	8782	8836	8891	8945	8999	9054	9108	9162	9217	
97	9271	9325	9380	9434	9488	9542	9597	9651	9705	9760	
98	9814	9868	9923	9977	0031	0085	0140	0194	0248	0303	
99	9030357	0411	0466	0520	0574	0628	0683	0737	0791	0846	
8000	0900	0954	1008	1063	1117	1171	1226	1280	1334	1388	
D. {54} P. {		5	11	16	22	27	32	38	43	49	

Thus, if the number consist of four figures only, we have simply to copy out the figures in the column headed 0, prefix a decimal point, and the proper characteristic.

Ex. Log 7991 = 3·9026011, log 7·995 = ·9028185.

When we speak of a number consisting of four figures only, we include such numbers as ·003654, ·07682, &c., the number of zeros immediately following the decimal points not being counted.

$$\text{Thus, log } ·07997 = \overline{2}·9029271$$
$$\log ·007992 = \overline{3}·9026555.$$

When the number contains five figures, as, for instance, 79936, we look along the line containing the first four figures —viz., 7993—of the number until the eye rests upon the column headed 6, the fifth figure. We then take the first three figures of the column headed 0, and affix the four figures of the column headed 6 in the horizontal line of the first four figures of the number.

$$\text{Thus, log } 79936 = 4·9027424$$
$$\log ·079927 = \overline{2}·9026935.$$

ALGEBRA.

It will be seen from the portion of the logarithmic table above extracted, that when the first three figures of the logarithm—viz., 902—have been once printed, they are not repeated, but must be understood to belong to every four figures in each column, until they are superseded by higher figures, as 903. When, however, this change is intended to be made at any place not at the commencement of a horizontal row, the first of the four figures corresponding to the change is usually printed either in different type, or, as above, with a bar over it. Thus we have above $\bar{0}031$, indicating that from this point we must prefix 903 instead of 902.

$$\text{Thus, log } 79{\cdot}986 = 1{\cdot}9030140,$$
$$\text{log } {\cdot}0079987 = \bar{3}{\cdot}9030194.$$

156. *To find the logarithm of a number not contained in the tables.*

Ex. Find the logarithm of 799·1635.

Since * the *mantissa* of the number 79916·35 is the same as the *mantissa* of the given number, and that the first five figures are contained in the tables, we may proceed as follows—

(1.) Take out from the tables the mantissa corresponding to the number 79916. This is ·9026337.

(2.) Take out the mantissa of the *next higher* number in the tables—viz., 79917. This is ·9026392.

(3.) Find the difference between these mantissæ. This is called the tabular difference, being the difference of the mantissæ for a difference of *unity* in the numbers. We find tab. diff. = ·0000055, which we call D.

(4.) Then *assuming*, as will be proved in Art. 166, that small differences in numbers are proportional to the differences of the corresponding logarithms, we find the difference for ·35′ = ·35 × ·0000055 = ·0000019, retaining only 7 figures. This is often called d.

(5.) Now adding this value of d to the mantissa for the number 79916, we get the mantissa corresponding to the number 79916·35.

* Thus log 79916·35 = log (100 × 799·1635) = 2 + log 799·1635.

THE USE OF TABLES.

(6.) Lastly, prefix to this mantissa the proper characteristic.

The whole operation may stand thus—

$$\text{M. of log } 79916 = \cdot 9026337 \quad \ldots\ldots\ldots\ldots \quad (1).$$
$$\text{M. of log } 79917 = \cdot 9026392$$
$$\text{Tabular difference or } D = \cdot 0000055$$

Hence, difference for ·35 or d
$$= \cdot 35 \times \cdot 0000055 = \cdot 0000019 \ldots\ldots\ldots\ldots\ldots (2).$$

Hence, adding (1.) and (2.)—
$$\text{M. of log } 79916 \cdot 35 = \cdot 9026356.$$
$$\therefore \log 799 \cdot 1635 = 2 \cdot 9026356.$$

or better thus, omitting the useless ciphers—

$$\text{M. of log } 79916 = \cdot 9026337$$
$$\text{M. of log } 79917 = \cdot 9026392$$
$$\therefore D = \quad\quad 55$$
$$\text{Hence, } d = \cdot 35 \times 55 = \quad 19$$
$$\therefore \text{M. of log } 79916 \cdot 35 = \cdot 9026356, \text{ as before.}$$

In the next article we shall show how the required difference may be obtained by inspection from the tables.

157. *Proportional parts.*

We saw in the example just worked that the tab. diff. (omitting the useless ciphers) is 55, and if we examine the table in Art. 155, we shall find the difference between the mantissæ of any two consecutive numbers there to be 54 or 55—generally 54. The number 54 is therefore placed in a separate column at the right of the table, and headed D.

The student will understand that the tab. diff. changes from time to time, and is not *always* 54 or 55.

Now assuming as in (4.) of the last article, we have—

Diff. for $\cdot 1 = 54 \times \cdot 1 = 5$ Diff. for $\cdot 6 = 54 \times \cdot 6 = 32$
,, $\cdot 2 = 54 \times \cdot 2 = 11$,, $\cdot 7 = 54 \times \cdot 7 = 38$
,, $\cdot 3 = 54 \times \cdot 3 = 16$,, $\cdot 8 = 54 \times \cdot 8 = 43$
,, $\cdot 4 = 54 \times \cdot 4 = 22$,, $\cdot 9 = 54 \times \cdot 9 = 49$
,, $\cdot 5 = 54 \times \cdot 5 = 27$

We find therefore the numbers 5, 11, 16, 22, 27, 32, 38, 43, 49 placed in a horizontal row at the bottom marked P; in the columns respectively headed 1, 2, 3, 4, &c.

ALGEBRA.

Hence, if we require the difference for (say) ·7, we take out the number 38 from the horizontal row marked P, instead of being at the trouble to find it by actual computation.

The following example will illustrate how we proceed when we require the difference for a decimal containing more than one decimal figure. No explanation is needed.

Ex. Find log 7994·3726—

$$\begin{aligned}
\text{M. of log } 79943 &= ·9027804 \\
\text{Diff. for } 7 &= 38 \\
\text{,, } 2 &= 1|1 \\
\text{,, } 6 &= |32 \\
\therefore \text{M. of log } 79943726 &= ·9027843
\end{aligned}$$

Hence, log 7994·3726 = 3·9027843.

158. *Having given the logarithm of a number to find the number.*

After the explanations of Art. 156, the method of working the following examples will be easily understood :—

Ex. 1. Find the number whose logarithm is $\bar{1}$·9030173. Taking from the tables the mantissæ next above and below, we have—

$$\begin{aligned}
·9030194 &= \text{M. of log } 79987 \\
·9030140 &= \text{M. of log } 79986 \quad\ldots\ldots\ldots\ldots(1). \\
\therefore \quad 54 &= \text{·D.} \\
\text{Again, } ·9030173 &= \text{M. of log N} \ldots\ldots\ldots\ldots\ldots(2).
\end{aligned}$$

Hence, subtracting (1) from (2)—

33 = d, the difference between the logarithms of the required number and the next lower.

Now $\dfrac{33}{54}$ = ·61, the difference between the next lower number and the required number.

Hence ·9030173 = M. of log 79986·61 ;

∴ $\bar{1}$·9030173 = log ·7998661 ;

∴ ·7998661 is the number required.

Ex. 2.* Find the value of $\dfrac{(1·023)^3 \times (·00123)^{\frac{1}{4}}}{(1·32756)^4}$.

---

* The logarithms used in this example are taken from the tables.

TRIGONOMETRICAL TABLES.

We have—
log N = 3 log 1·023 + ¼ log ·00123 − 4 log 1·32756.
Now, 3 log 1·023 = 3 × ·0098756   = ·0276268
¼ log ·00123 = ¼ ($\bar{3}$·0899051)
   = ¼ ($\bar{4}$ + 1·0899051)   = $\bar{1}$·2724763
∴ adding, 3 log. 1·023 + ¼ log ·00123 = $\bar{1}$·3001031
Again, M. of log 13275 = ·1230345
and diff. for   ·6 =   196
∴ M. of log 13275·6 = ·1230541
∴ 4 log 1·32756 = 4 × ·1230541   = ·4922164
Then, subtracting, log N   = $\bar{2}$·5078867
Hence we have, ·5078867 = M. of log N,
           and ·5078828 = M. of log 32202;
   ∴   39 = d,
   also 135 = D,
   and $\frac{39}{135}$ = ·29.
∴ ·5078867 = M. of log 32202·29;
∴ $\bar{2}$·5078867 = log ·03220229.
Hence ·03220229 is the number required.

Ex. XXXV.

1. Find the logarithm to base 4 of the following numbers:
16, 64, 2, ·25, ·0625, 8.
2. Find the value of $\log_8 32$, $\log_{\sqrt[3]{5}} 25$, $\log_{81} ·729$.
3. Given log 2 = ·3010300, and log 3 = ·4771213, find the logarithms of 12, 36, 45, 75, ·04, 3·75, ·$\dot{6}$, ·07$\dot{4}$.
4. Given log 20763 = 4·3172901, what is the logarithm of 2·0763, 2076·3, ·020763, ·0020763?
5. Write down the characteristics of the common logarithms of 29·6, ·25402, ·0034, 6176·003.
6. Given log 20·912 = 1·3203956, what numbers correspond to the following logarithms:—$\bar{2}$·3203956, 6·3203956, $\bar{1}$·3203956, 4·3203956?

ALGEBRA.

7. Given log 20·713 = 1·3162430, and log 20714 = 3·3162640, find log ·2071457.

8. Given log 3·4937 = ·5432856, and log 3·4938 = ·5432980, find the number whose logarithm is 3·5432930.

9. Given log 1·05 = ·0211893, log 2·7 = 1·4313638, log 135 = 2·1303338, find the value of $\log \dfrac{(2\cdot7)^{\frac{1}{3}} \times \sqrt[5]{13\cdot5}}{(1\cdot05)^6}$.

10. Given log 18 = 1·2552725, and log 2·4 = ·3802112, find the value of log ·00135.

11. What are the characteristics of $\log_3 1167$, and $\log_4 1965$?

12. Having log 2 = ·3010300, and log 3 = ·4771213, find x when $18^x = 125$.

13. Given log 47582 = 4·6774427, and log 47583 = 4·6774518, find log 47·58275.

14. Given log 5·2404 = ·7193644, and log 524·05 = 2·7193727, find log 5240463.

15. Given log ·56145 = $\overline{1}$·7493111, and log 56·146 = 1·7493188, find log $\sqrt[7]{·05614581}$.

16. Given log 61683 = 4·7901655, and log 616·84 = 2·7901725, find the number whose logarithm is $\overline{2}$·7901693.

17. Find the value of $(1·05)^{15}$, having given log 1·05 = ·0211893, log 20789 = 4·3178336, and log 20790 = 4·3178545.

18. Find the compound interest of £120 for 10 years at 4 per cent. per annum, having given log 1·04 = ·0170333, log 14802 = 4·1703204, and log 14803 = 4·1703497.

19. A corporation borrows £8,630 at 4½ per cent. compound interest, what annual payment will clear off the debt in 20 years?

Log 1·045 = ·0191163, log 4·1464 = 6176712, and
log 4·1465 = ·6176817.

Exponential Theorem.

159. *To expand a^x in a series of ascending powers of* x.

$$a^x = \{1 + (a - 1)\}^x$$
$$= 1 + x(a - 1) + \frac{x(x - 1)}{1 \cdot 2}(a - 1)^2$$
$$+ \frac{x(x - 1)(x - 2)}{1 \cdot 2 \cdot 3}(a - 1)^3 + \&c.$$
$$= 1 + x(a - 1) + \frac{x^2 - x}{2}(a - 1)^2$$
$$+ \frac{x^3 - 3x^2 + 2x}{6}(a - 1)^3 + \&c.$$

Collecting the terms involving the first power only of x, we have
$$a^x = 1 + \{(a - 1) - \tfrac{1}{2}(a - 1)^2 + \tfrac{1}{3}(a - 1)^3 - \&c.\} x$$
$$+ \text{terms in } x^2, x^3, \&c.$$

Assume then
$$a^x = 1 + Ax + Bx^2 + Cx^3 + \&c. \ldots\ldots\ldots (1),$$
Where $A = (a - 1) - \tfrac{1}{2}(a - 1)^2 + \tfrac{1}{3}(a - 1)^3 - \&c.$ (2).
From (1), squaring and arranging, we have
$$a^{2x} = 1 + 2Ax + (A^2 + 2B)x^2 + (2C + 2AB)x^3 + \&c\ldots(3).$$
But from (1), putting therein $2x$ for x, we also get
$$a^{2x} = 1 + A(2x) + B(2x)^2 + C(2x)^3 + \&c.$$
$$= 1 + 2Ax + 4Bx^2 + 8Cx^3 + \&c. \ldots\ldots(4).$$

Hence the series in (3) and (4) are identical, and we may therefore equate the coefficients of like powers of x.

Hence $4B = A^2 + 2B$, or $2B = A^2$. Therefore $B = \dfrac{A^2}{1 \cdot 2}$.

$8C = 2C + 2AB$, or $6C = 2A \cdot \dfrac{A^2}{1.2}$. $\therefore C = \dfrac{A^3}{1.2.3}$. &c.

Hence, substituting in (1),
$$a^x = 1 + \frac{Ax}{1} + \frac{A^2 x^2}{1.2} + \frac{A^3 x^3}{1.2.3} + \&c\ldots\ldots\ldots(5).$$

Since this equation is true for all values of x, we may put
$$Ax = 1, \text{ and } \therefore x = \frac{1}{A}.$$

We then have $a^{\frac{1}{A}} = 1 + \dfrac{1}{1} + \dfrac{1}{1.2} + \dfrac{1}{1.2.3} + $ &c.

Hence, by Art. 145, Ex. 5,
$$a^{\frac{1}{A}} = e, \text{ or } a = e^A \ldots\ldots\ldots\ldots\ldots\ldots(6).$$

Taking the logarithm of each side of the equation to base e, we get
$$A = \log_e a \ldots\ldots\ldots\ldots\ldots\ldots(7).$$

Hence, substituting in (5) for this value of A, we get
$$a^x = 1 + \frac{x \log_e a}{1} + \frac{(x \log_e a)^2}{1.2} + \frac{(x \log_e a)^3}{1.2.3} + \&c.$$

The $(n + 1)$th term of this series is $\dfrac{(x \log_e a)^n}{\lfloor n}$, and the formula itself is known by the name of the **Exponential Theorem**.

Cor. 1. Put $a = e$, then since $\log_e e = 1$, we have
$$e^x = 1 + \frac{x}{1} + \frac{x^2}{1.2} + \frac{x^3}{1.2.3} + \&c.$$

Cor. 2. Since in (7) we have $A = \log_e a$, it follows at once from (2) that
$$\log_e a = (a - 1) - \tfrac{1}{2}(a - 1)^2 + \tfrac{1}{3}(a - 1)^3 - \&c.$$

Logarithmic Series.

160. *To show that* $\log_e (1 + x) = x - \dfrac{1}{2} x^2 + \dfrac{1}{3} x^3 - $ &c.

and that $\log_e (1 - x) = -x - \dfrac{1}{2} x^2 - \dfrac{1}{3} x^3 - $ &c.

LOGARITHMIC SERIES.

In Art. 159, Cor. 1, we have
$$\log_e a = (a - 1) - \frac{1}{2}(a - 1)^2 + \frac{1}{3}(a - 1)^3 - \&c.$$

Put $1 + x$ for a, and therefore x for $a - 1$, then
$$\log_e (1 + x) = x - \frac{1}{2}x^2 + \frac{1}{3}x^3 - \&c. \ldots\ldots\ldots\ldots(1).$$

Now put $- x$, then
$$\log_e (1 - x) = - x - \frac{1}{2}x^2 - \frac{1}{3}x^3 - \&c. \ldots\ldots\ldots(2).$$

These series are not suitable for the calculation of logarithms. We proceed to obtain from them more convergent series.

161. *To show that* $\log_e (n + 1)$
$$= \log_e n + 2 \left\{ \frac{1}{2n+1} + \frac{1}{3} \cdot \frac{1}{(2n+1)^3} + \frac{1}{5} \cdot \frac{1}{(2n+1)^5} + \&c. \right\}.$$

Now $\log_e \dfrac{1 + x}{1 - x} = \log_e (1 + x) - \log_e (1 - x)$

$$= \left(x - \frac{1}{2}x^2 + \frac{1}{3}x^3 - \frac{1}{4}x^4 + \frac{1}{5}x^5 - \&c. \right)$$
$$- \left(- x - \frac{1}{2}x^2 - \frac{1}{3}x^3 - \frac{1}{4}x^4 - \frac{1}{5}x^5 - \&c. \right)$$
$$= 2 \left(x + \frac{1}{3}x^3 + \frac{1}{5}x^5 + \&c. \right) \ldots\ldots\ldots\ldots\ldots\ldots(1).$$

Since this is true for all values of x, we may put
$$\frac{1 + x}{1 - x} = \frac{m}{n}, \text{ and therefore also } x = \frac{m - n}{m + n}.$$

We then have $\log_e \dfrac{m}{n}$
$$= 2 \left\{ \frac{m - n}{m + n} + \frac{1}{3} \left(\frac{m - n}{m + n} \right)^3 + \frac{1}{5} \left(\frac{m - n}{m + n} \right)^5 + \&c. \right\} (2).$$

Now put $m = n + 1$, and therefore
$$m - n = 1, \text{ and } m + n = 2n + 1;$$

ALGEBRA.

Then we have $\log_e \frac{n+1}{n}$
$$= 2\left\{\frac{1}{2n+1} + \frac{1}{3}\cdot\frac{1}{(2n+1)^3} + \frac{1}{5}\cdot\frac{1}{(2n+1)^5} + \&c.\right\}.$$

Hence, transposing, $\log_e (n+1)$
$$= \log_e n + 2\left\{\frac{1}{2n+1} + \frac{1}{3}\cdot\frac{1}{(2n+1)^3} + \frac{1}{5}\cdot\frac{1}{(2n+1)^5} + \&c.\right\};$$

an important series, by which the logarithm of any number may be easily computed when that of the next lower is known.

162. *To show that* $\log_e (x+1)$
$$= 2\log_e x - \log_e (x-1) - 2\left\{\frac{1}{2x^2-1} + \frac{1}{3}\cdot\frac{1}{(2x^2-1)^3} + \&c.\right\}.$$

We have in (2) of the last Art., putting
$m = x^2$, and $n = x^2 - 1$, and therefore
$m - n = 1$, and $m + n = 2x^2 - 1$,

$$\log_e \frac{x^2}{x^2-1} = 2\left\{\frac{1}{2x^2-1} + \frac{1}{3}\cdot\frac{1}{(2x^2-1)^3} + \&c.\right\}.$$

But $\log_e \dfrac{x^2}{x^2-1} = \log_e \dfrac{x^2}{(x+1)(x-1)}$
$$= 2\log_e x - \log_e (x+1) - (x-1).$$

Hence $2\log_e x - \log_e (x+1) - \log_e (x-1)$
$$= 2\left\{\frac{1}{2x^2-1} + \frac{1}{3}\cdot\frac{1}{(2x^2-1)^3} + \&c.\right\}.$$

Or, transposing, $\log_e (x+1)$
$$= 2\log_e x - \log_e (x-1) - 2\left\{\frac{1}{2x^2-1} + \frac{1}{3}\cdot\frac{1}{(2x^2-1)^3} + \&c.\right\}.$$

This is a very useful formula, by which the logarithm of any number is found from the logarithms of the two numbers next preceding.

163. *To show that, when* x *is large, whatever be the base,*
$$2\log x = \log(x+1) + \log(x-1) + \frac{1}{2x}\left\{1 + \frac{1}{6x^2} + \frac{1}{90x^4} + \&c.\right\}\log\frac{x+1}{x-1}.$$

LOGARITHMIC SERIES.

Now, $\log_e \dfrac{x^2}{x^2 - 1} = -\log_e \dfrac{x^2 - 1}{x^2} = -\log_e \left(1 - \dfrac{1}{x^2}\right)$

$= -\left(-\dfrac{1}{x^2} - \dfrac{1}{2} \cdot \dfrac{1}{x^4} - \dfrac{1}{3} \cdot \dfrac{1}{x^6} - \&c.\right)$

$= \dfrac{1}{x^2} + \dfrac{1}{2} \cdot \dfrac{1}{x^4} + \dfrac{1}{3} \cdot \dfrac{1}{x^6} \ldots\ldots\ldots\ldots\ldots(1).$

And $\log_e \dfrac{x+1}{x-1} = \log_e \dfrac{1 + \dfrac{1}{x}}{1 - \dfrac{1}{x}}$, and therefore, by Art. 161 (1),

$= 2\left\{\dfrac{1}{x} + \dfrac{1}{3} \cdot \dfrac{1}{x^3} + \dfrac{1}{5} \cdot \dfrac{1}{x^5} + \&c.\right\} \ldots\ldots(2).$

Dividing (1) by (2), we have, after performing the division,

$\log_e \dfrac{x^2}{x^2 - 1} \div \log_e \dfrac{x+1}{x-1} = \dfrac{1}{2}\left\{\dfrac{1}{x} + \dfrac{1}{6 x^3} + \dfrac{7}{90 x^5} + \&c.\right\} \ldots (3).$

Now, $\log_e \dfrac{x^2}{x^2 - 1} = 2\log_e x - \log_e (x + 1) - \log_e (x - 1)$;

Hence, multiplying each side of (3) by $\log_e \dfrac{x+1}{x-1}$, and transposing, &c.

$2\log_e x = \log_e (x + 1) + \log_e (x - 1)$
$+ \dfrac{1}{2x}\left\{1 + \dfrac{1}{6 x^2} + \dfrac{7}{90 x^4} + \&c.\right\}\log_e \dfrac{x+1}{x-1}\ldots\ldots(4).$

Now it is shown (Art. 162) that

$\log_e x = \log_e a \cdot \log_a x.$

Hence we may replace the base e in (4) by the base a, if we introduce the factor $\log_e a$ into every term. Dividing each side of the newly-formed equation by this factor, we then get the formula in (4) expressed thus:

$2\log_a x = \log_a (x + 1) + \log_a (x - 1)$
$+ \dfrac{1}{2x}\left\{1 + \dfrac{1}{6 x^2} + \dfrac{7}{90 x^4} + \&c.\right\}\log_a \dfrac{x+1}{x-1}.$

ALGEBRA.

Logarithms to base e are called Napierian logarithms, from their inventor.

It is evident that, by the series of the last articles, and by similar series, Napierian logarithms may be calculated. It will be necessary to use each series however for prime numbers only, since composite numbers may be broken up into their prime factors, and their logarithms easily obtained from those of their factors.

Ex. 1. Find $\log_e 2$, $\log_e 3$.

We have

$$\log_e \frac{m}{n} = 2\left\{\frac{m-n}{m+n} + \frac{1}{3}\left(\frac{m-n}{m+n}\right)^3 + \frac{1}{5}\left(\frac{m-n}{m+n}\right)^5 + \&c.\right\}.$$

Put $m = 2$, $n = 1$, and $\therefore \frac{m-n}{m+n} = \frac{2-1}{2+1} = \frac{1}{3}$.

We then have

$$\log_e 2 = 2\left\{\frac{1}{3} + \frac{1}{3}\left(\frac{1}{3}\right)^3 + \frac{1}{5}\left(\frac{1}{3}\right)^5 + \&c.\right\};$$

Or, reducing to decimals and adding
$$= \cdot6931471\ldots$$

Similarly, putting $m = 3$, $n = 2$, and $\therefore \frac{m-n}{m+n} = \frac{1}{5}$, we have

$$\log_e \frac{3}{2} = 2\left\{\frac{1}{5} + \frac{1}{3}\left(\frac{1}{5}\right)^3 + \frac{1}{5}\left(\frac{1}{5}\right)^5 + \&c.\right\};$$

or, since $\log_e \frac{3}{2} = \log_e 3 - \log_e 2$,

we have,

$$\log_e 3 = \log_e 2 + 2\left\{\frac{1}{5} + \frac{1}{3}\left(\frac{1}{5}\right)^3 + \frac{1}{5}\left(\frac{1}{5}\right)^5 + \&c.\right\};$$

or, performing the necessary reductions, &c.,
$$= 1 \cdot 0986122\ldots$$

Ex. 2. Find $\log_e 5$, $\log_e 10$.

Put $m = 5$, $n = 4$, and $\therefore \frac{m-n}{m+n} = \frac{1}{9}$, then since

$\log_e \frac{5}{4} = \log_e 5 - 2\log_e 2$, we have, by transposing,

$$\log_e 5 = 2\log_e 2 + 2\left\{\frac{1}{9} + \frac{1}{3}\left(\frac{1}{9}\right)^3 + \frac{1}{5}\left(\frac{1}{9}\right)^5 + \&c.\right\};$$

or, reducing, &c.,
$$= 1\cdot 6094379\ldots$$

Again,
$$\log_e 10 = \log_e(2 \times 5) = \log_e 2 + \log_e 5$$
$$= \cdot 6931471\ldots \quad + 1\cdot 6094379\ldots$$
$$= 2\cdot 3025850\ldots$$

Ex. 3. Find $\log_e 16205$.

$$\log_e 16205 = \log_e(2^3 \times 3^4 \times 5^2 + 5)$$
$$= \log_e\left\{2^3 \times 3^4 \times 5^2\left(1 + \frac{1}{2^3 \times 3^4 \times 5}\right)\right\}$$
$$= 3\log_e 2 + 4\log_e 3 + 2\log_e 5 + \log_e\left(1 + \frac{1}{2^3 \times 3^4 \times 5}\right).$$

By substituting the values of $\log_e 2$, $\log_e 3$, $\log_e 5$, already found, and by developing $\log_e\left(1 + \frac{1}{2^3 \times 3^4 \times 5}\right)$, the required logarithm is known.

Calculation of Common Logarithms.

164. To show that $\log_b N = \frac{1}{\log_a b} \cdot \log_a N$.

Let $\qquad N = a^x = b^y$ ....(1).

Then we have, from the definition of a logarithm,
$$x = \log_a N\ldots\ldots\ldots\ldots\ldots\ldots\ldots(2).$$
$$y = \log_b N\ldots\ldots\ldots\ldots\ldots\ldots\ldots(3).$$

Again, we have in (1), $a^x = b^y$, or taking logarithms to base a,
$$x = y\log_a b:$$

or, substituting from (2) and (3),
$$\log_a N = \log_b N \cdot \log_a b$$
$$\therefore \log_b N = \frac{1}{\log_a b} \cdot \log_a N.$$

Cor. Put $b = 10$, and $a = e$, then
$$\log_{10} N = \frac{1}{\log_e 10} \cdot \log_e N.$$

Hence common logarithms may be found from Napierian logarithms by multiplying the latter by $\frac{1}{\log_e 10}$.

We call this quantity the *modulus* of the common system of logarithms; and if we represent the modulus by μ, we have
$$\mu = \frac{1}{\log_e 10} = \frac{1}{2\cdot 30258509\ldots} = \cdot 43429448\ldots$$

165. By multiplying by μ the series obtained in Arts. 160, 161, and putting $\mu \log_e = \log_{10}$, we have, omitting the suffix 10.

$$\log(1 + x) = \mu \left\{ x - \frac{1}{2}x^2 + \frac{1}{3}x^3 - \&c. \right\} \ldots (1).$$

$$\log(1 - x) = \mu \left\{ -x - \frac{1}{2}x^2 - \frac{1}{3}x^3 - \&c. \right\} \ldots (2).$$

$$\log \frac{1+x}{1-x} = 2\mu \left\{ x + \frac{1}{3}x^3 + \frac{1}{5}x^5 + \&c. \right\} \ldots (3).$$

$$\log \frac{m}{n} = 2\mu \left\{ \frac{m-n}{m+n} + \frac{1}{3}\left(\frac{m-n}{m+n}\right)^3 + \frac{1}{5}\left(\frac{m-n}{m+n}\right)^5 + \&c. \right\} (4).$$

$$\log(n+1)$$
$$= \log n + 2\mu \left\{ \frac{1}{2n+1} + \frac{1}{3}\left(\frac{1}{2n+1}\right)^3 + \frac{1}{5}\left(\frac{1}{2n+1}\right)^5 + \&c. \right\} (5).$$

$$\log(x+1)$$
$$= 2\log x - \log(x-1) - 2\mu \left\{ \frac{1}{2x^2-1} + \frac{1}{3}\left(\frac{1}{2x^2+1}\right)^2 + \&c. \right\} (6).$$

Theory of Proportional Parts.

166. *To find the logarithm of a number containing* $(n+1)$ *digits, when the table contains the logarithms of numbers containing* n *digits.*

We have

$$\log(n+\delta) - \log n = \log\frac{n+\delta}{n} = \log\left(1 + \frac{\delta}{n}\right)$$

$$= \mu\left(\frac{\delta}{n} - \frac{1}{2}\cdot\frac{\delta^2}{n^2} + \frac{1}{3}\cdot\frac{\delta^3}{n^3} - \&c.\right).$$

Suppose n to be an integer containing 5 figures, and δ a quantity less than unity.

We then have, since $\mu\ (= \cdot 434...)$ is $< \frac{1}{2}$,

$$\mu\cdot\frac{\delta^2}{2\,n^2} < \frac{1}{4}\cdot\frac{1}{10000^2} \text{ and } \therefore < \cdot 000000003;$$

$$\mu\cdot\frac{\delta^3}{3\,n^3} < \frac{1}{6}\cdot\frac{1}{10000^3} \text{ and } \therefore < \cdot 0000000000002, \&c.$$

Hence, at least as far as the *seventh* place of decimals, the omission of all the terms of the above series after the first will not affect the result.

We then have

$$\log(n+\delta) - \log n = \mu\cdot\frac{\delta}{n}\dots\dots\dots\dots\dots\dots(1).$$

And, similarly,

$$\log(n+1) - \log n = \mu\cdot\frac{1}{n} = d \text{ suppose,}\dots\dots(2).$$

Then we have,

$$\log(n+\delta) - \log n = \delta d\dots\dots\dots\dots\dots(3).$$

Now d is the difference between the logarithms of two consecutive numbers;

And δ is the difference (less than unity) of two numbers, the logarithm of the greater of which is required.

Hence we have the following rule:

To find the logarithm of a number which lies between two consecutive numbers, multiply the difference of the given

ALGEBRA.

number and the next lower in the tables by the difference of the logarithms of the numbers next above and below the given number, and add this result to the logarithm of the smaller given number.

This rule has been assumed in Chapter XVIII., to which the student is referred for practical applications.

Remark. Since the mantissæ of all numbers, having the same digits, are the same, it follows that the above applies equally to numbers the integral part of which contains more than five digits.

167. *To prove that*

$$n^r - n(n-1)^r + \frac{n(n-1)}{\lfloor 2}(n-2)^r - \frac{n(n-1)(n-2)}{\lfloor 3}(n-3)^r + \&c. = \lfloor n, \text{ if } r = n, \text{ and } = 0, \text{ if } r < n.$$

We have

$$(e^x - 1)^n = e^{nx} - n \cdot e^{(n-1)x} + \frac{n(n-1)}{\lfloor 2} e^{(n-2)x} - \frac{n(n-1)(n-2)}{\lfloor 3} e^{(n-3)x} + \&c.$$

If we now expand each of the quantities $e^{nx}, e^{(n-1)x}$, &c., by the Exponential Theorem, and write down the coefficient of x^r in the whole expression, we have

Coefficient of x^r in $(e^x - 1)^n$

$$= \frac{n^r}{\lfloor r} - n \cdot \frac{(n-1)^r}{\lfloor r} + \frac{n(n-1)}{\lfloor 2} \cdot \frac{(n-2)^r}{\lfloor r}$$
$$- \frac{n(n-1)(n-2)}{\lfloor 3} \cdot \frac{(n-3)^r}{\lfloor r} + \&c. \quad\quad\quad (1).$$

Again, $(e^x - 1)^n$

$$= \left(x + \frac{x^2}{\lfloor 2} + \frac{x^3}{\lfloor 3} + \&c.\right)^n$$

$$= x^n + \text{terms involving higher powers of } x \quad\quad\quad (2).$$

Hence the coefficient of x^n is *unity*; and since the expansion contains no terms with lower powers of x than the nth, we may say that

PROPORTIONAL PARTS.

The coefficient of x^r in $(e^x - 1)^n = 1$, when $r = n$, and $= 0$, when $r < n$, $\}\ldots(3)$.

Hence from (1) and (3), equating the coefficients of x^r,

$$\frac{n^r}{\lfloor r} - n\frac{(n-1)^r}{\lfloor r} + \frac{n(n-1)}{\lfloor 2} \cdot \frac{(n-2)^r}{\lfloor r}$$
$$- \frac{n(n-1)(n-2)}{\lfloor 3} \cdot \frac{(n-3)^r}{\lfloor r} + \&c.$$
$$= 1, \text{ when } r = n,$$
$$\text{and} = 0, \text{ when } r < n.$$

$$\therefore n^r - n(n-1)^r + \frac{n(n-1)}{\lfloor 2}(n-2)^r$$
$$- \frac{n(n-1)(n-2)}{\lfloor 3}(n-3)^r + \&c.$$
$$= \lfloor n, \text{ if } r = n,$$
$$\text{and} = 0, \text{ if } r < n.$$

Ex. XXXVI.

Show that

1. $\dfrac{e^{2x} - 1}{xe^x} = 2\left\{1 + \dfrac{x^2}{\lfloor 3} + \dfrac{x^4}{\lfloor 5} + \&c.\right\}$

2. $\dfrac{e^{2x} + 1}{e^x} = 2\left\{1 + \dfrac{x^2}{\lfloor 2} + \dfrac{x^4}{\lfloor 4} + \&c.\right\}$

3. $m^n = m + \dfrac{m(m-1)}{1.2}(2^n - 2)$
$+ \dfrac{m(m-1)(m-2)}{1.2.3}(3^n - 3.2^n + 3) + \&c.$

4. $n - n(n-1) + \dfrac{n(n-1)(n-2)}{2} - \&c. = 0.$

5. $n^{n-1} - n(n-1)^{n-1} + \dfrac{n(n-1)}{1.2}(n-2)^{n-1} - \&c. = 0.$

6. $n^n - n(n-1)^n + \dfrac{n(n-1)}{1.2}(n-2)^n - \&c. = \lfloor n.$

7. $n^{n+1} - n(n-1)^{n+1} + \dfrac{n(n-1)}{1 \cdot 2}(n-2)^{n+1} - \&c.$
$= \dfrac{1}{2}n \lfloor n+1.$

8. $1^n - n \cdot 2^n + \dfrac{n(n-1)}{1 \cdot 2}3^n - \&c. + (-1)^n \cdot (n+1)^n$
$= (-1)^n \lfloor n.$

9. $1^m + 2^m + 3^m + \&c. + n^m = \dfrac{n^{m+1}}{m+1} + \dfrac{n}{2} + m \cdot \dfrac{n^{m-1}}{1 \cdot 2} + \&c.$

10. If p, q are each less than 1, show that $\dfrac{\log(1-p)}{\log(1-q)}$ is less than $\dfrac{p}{q - pq}$, and greater than $\dfrac{p - pq}{q}$.

11. If $u_r = r^p$, show that
$u_1 - nu_2 + \dfrac{n(n-1)}{1 \cdot 2}u_3 + \&c. + (-1)^n u_{n+1} = (-1)^n \lfloor n.$

12. Show that the limit of $\left(1 + \dfrac{1}{nx}\right)^x$, when x is infinite, is $e^{\frac{1}{n}}$.

13. Having given that $1 \cdot 2 \cdot 3 \ldots n = \sqrt{2\pi} \cdot n^{n+\frac{1}{2}} \cdot e^{-n}$, when n is increased without limit, show that
$$1 \cdot 3 \cdot 5 \ldots (2n-1) = 2^{n+\frac{1}{2}} \cdot n^n \cdot e^{-n},$$
when n is increased without limit.

14. Hence show that the limit of
$$\dfrac{1 \cdot 3 \cdot 5 \ldots (2n-1)}{2 \cdot 4 \cdot 6 \ldots 2n} \cdot \dfrac{1}{2n-1} \text{ is } n^{-\frac{3}{2}},$$
when n is increased without limit.

15. Show that
$$\left(1 + \dfrac{x}{2} - \dfrac{x^2}{12} + \dfrac{x^3}{24} - \dfrac{19 x^4}{720} + \&c.\right) \log_e(1+x) = x.$$

16. If $a = b - x$, show that $\log_e a$
$= \log_e b - \dfrac{1}{2b-x}\left\{\left(2-\dfrac{1}{2}\right)x + \left(\dfrac{2}{2}-\dfrac{1}{3}\right)\dfrac{x^2}{b} + \left(\dfrac{2}{3}-\dfrac{1}{4}\right)\dfrac{x^3}{b^2} + \&c. \right\}.$

THE MULTINOMIAL THEOREM

17. If $a^2 + 1 = 0$, show that

(1.) $e^{ax} + e^{-ax} = 2\left\{1 + \dfrac{x^2}{\lfloor 2} + \dfrac{x^4}{\lfloor 4} + \&c.\right\}$

(2.) $e^{ax} - e^{-ax} = 2a\left\{x + \dfrac{x^3}{\lfloor 3} + \dfrac{x^5}{\lfloor 5} + \&c.\right\}$

18. If a, b, c are consecutive numbers, show that

$$\log b = \frac{1}{2}(\log a + \log c) + \mu\left\{\frac{1}{2ac+1} + \frac{1}{3}\cdot\frac{1}{(2ac+1)^3} + \frac{1}{5}\frac{1}{(2ac+1)^5} + \&c.\right\}$$

XXXV.—
1. 2, 3, ½, − 1, − 2, 1·5.   2. ⅔, 6, 1·5,

ALGEBRA.

3. 1·0791812, 1·5563025, 1·6532125, 1·8750613, $\bar{2}$·6020600, ·5740313, $\bar{1}$·8239087, $\bar{2}$·8696761.
4. 1·3172901, 3·3172901, $\bar{2}$·3172901, $\bar{3}$·3172901.
5. 1, $\bar{1}$, $\bar{3}$, $\bar{3}$.
6. ·020912, 2091200, ·20912, 20912.   7. $\bar{1}$·3162760.
8. 3493·768.   9. ·2427189.   10. 3·1303338.
11. 6, 5.   12. 1·67.   13. 1·6774495.
14. 6·7193696.   15. $\bar{1}$·8213310.   16. ·0616835.
17. 2·07892.   18. £177·63 nearly.   19. £663·449.
20. 2620·248.

LOGARITHMS.

332. The **Logarithm** of a number is the exponent by which a fixed *base* must be affected in order to be equal to the number. That is, if $a^x = N$, x is the logarithm of N to the base a, which is written $x = \log_a N$.

Thus, since $\quad 3^2 = 9, \quad 2 = \log_3 9.$
Since $\quad\quad\;\; 2^4 = 16, \quad 4 = \log_2 16.$
Since $\quad\quad\;\; 10^1 = 10, \; 10^2 = 100, \; 10^3 = 1000, \cdots,$

the positive numbers 1, 2, 3, $\cdots$, are respectively the logarithms of 10, 100, 1000, $\cdots$, to the base 10.

To the base 10 the logarithms of all numbers between 1 and 10, 10 and 100, 100 and 1000, $\cdots$, are incommensurable.

Since $\quad\quad\;\; 3^{-2} = \tfrac{1}{9}, \quad -2 = \log_3 \tfrac{1}{9}.$
Since $\quad\quad\;\; 10^{-1} = 0.1, \; 10^{-2} = 0.01, \; 10^{-3} = 0.001, \cdots,$

the negative numbers $-1, -2, -3, \cdots$, are respectively the logarithms of 0.1, 0.01, 0.001, $\cdots$, to the base 10.

333. Any positive number except 1 may evidently be taken as the base of logarithms. The following are the *fundamental properties* of logarithms to any base.

334. *The logarithm of 1 is zero.*

For $\quad\quad a^0 = 1. \quad \therefore \; \log_a 1 = 0.$

335. *The logarithm of the base itself is 1.*

For $\quad\quad a^1 = a. \quad \therefore \; \log_a a = 1.$

322

LOGARITHMS.

336. *The logarithm of a product equals the sum of the logarithms of its factors.*

For let $M = a^x$, $N = a^y$;

then $M \times N = a^{x+y}$.

Hence $\log_a(M \times N) = x + y = \log_a M + \log_a N$.

337. *The logarithm of a quotient equals the logarithm of the dividend minus that of the divisor.*

For let $M = a^x$, $N = a^y$;

then $M \div N = a^{x-y}$.

Hence $\log_a(M \div N) = x - y = \log_a M - \log_a N$.

338. *The logarithm of a positive number affected with any exponent equals the logarithm of the number multiplied by the exponent.*

Let $M = a^x$;

then, whatever be the value of p,

$$M^p = a^{px}.$$

Hence $\log_a(M^p) = px = p \log_a M$.

339. By § 338, the logarithm of any power of a number equals the logarithm of the number multiplied by the exponent of the power; and the logarithm of any root of a number equals the logarithm of the number divided by the index of the root.

340. From the principles proved above, we see that by the use of logarithms the operations of multiplication and division may be replaced by those of addition and subtraction, and the operations of involution and evolution by those of multiplication and division.

LOGARITHMS.

EXAMPLE. Find $\log_a \dfrac{\sqrt[3]{x^2}}{b^3 c^{\frac{5}{4}}}$ in terms of $\log_a x$, $\log_a b$, $\log_a c$.

$$\log_a \dfrac{\sqrt[3]{x^2}}{b^3 c^{\frac{5}{4}}} = \log_a x^{\frac{2}{3}} - \log_a (b^3 c^{\frac{5}{4}}) \qquad \S\ 336$$

$$= \log_a x^{\frac{2}{3}} - (\log_a b^3 + \log_a c^{\frac{5}{4}}) \qquad \S\ 335$$

$$= \tfrac{2}{3} \log_a x - 3 \log_a b - \tfrac{5}{4} \log_a c. \qquad \S\ 337$$

Exercise 99.

1. Find $\log_2 8$; $\log_4 64$; $\log_3 81$; $\log_2 32$; $\log_5 125$; $\log_4 \tfrac{1}{16}$; $\log_9 \tfrac{1}{81}$; $\log_3 \tfrac{1}{81}$; $\log_2 \tfrac{1}{32}$; $\log_3 \tfrac{1}{27}$; $\log_5 \tfrac{1}{125}$.

2. If 10 is the base, between what integral numbers does the logarithm of any number between 1 and 10 lie? Of any number between 10 and 100? Of any number between 100 and 1000? Of any number between 0.1 and 1? Of any number between 0.01 and 0.1? Of any number between 0.001 and 0.01?

In the next six examples express $\log_a y$ in terms of $\log_a b$, $\log_a c$, $\log_a x$, and $\log_a z$.

3. $y = x^3 b^{\frac{2}{3}} c^5$.
4. $y = \sqrt[3]{z^2} \cdot \sqrt[5]{c^6}$.
5. $y = \dfrac{b^3 x^2 z}{c^{\frac{2}{3}}}$.
6. $y = \sqrt[7]{b^2 x^4 z^3}$.
7. $y = \sqrt{xz^5} \cdot \sqrt[3]{b^2 c^7}$.
8. $y = \dfrac{\sqrt{x^3 z b^5}}{\sqrt[3]{x^2 z^4 c^5}}$.

COMMON LOGARITHMS.

341. The logarithms used for numerical computations are always those to base 10; for this reason logarithms to base 10 are called **Common Logarithms**. Hereafter in this chapter when no base is written, the base 10 is to be understood. It is evident that the logarithms of most numbers consist of an integral part and a fractional part.

LOGARITHMS.

The common logarithm of a number less than 1 is negative; but for convenience it is always written in such a form that the *fractional part* will be *positive*.

Thus the common logarithm of any number between

    1 and   10 will be   0 + a fraction,
   10 and   100 will be   1 + a fraction,
  100 and 1000 will be   2 + a fraction,
  0.1 and   1 will be -1 + a fraction,
  0.01 and 0.1 will be -2 + a fraction,
0.001 and 0.01 will be -3 + a fraction.

342. The integral part of a logarithm is called the **Characteristic**; and the fractional part, which is always *positive*, the **Mantissa**.

Thus in the logarithm 2.5798, 2 is the characteristic and .5798 the mantissa. In the logarithm $-2 + .7168$, which is often written $\bar{2}.7168$, -2 is the characteristic and .7168 the mantissa.

343. *If two numbers differ only in the position of the decimal point, their common logarithms have the same mantissa.*

For $\log (N \times 10^m) = \log N + m$.

Now if m is any whole number either positive or negative, the numbers N and $N \times 10^m$ will differ only in the position of the decimal point; and $\log N$ and $\log (N \times 10^m)$ will differ only by the whole number m; that is, they will have the same mantissa.

Thus having given that $\log 31 = 1.4914$, we have
$$\log 0.031 = \log (31 \div 1000) = \log 31 - \log 1000$$
$$= 1.4914 - 3$$
$$= 8.4914 - 10.$$

Here the negative characteristic -2 is written in the very common form $8-10$.

LOGARITHMS.

344. The *characteristic* of the common logarithm of any number can be determined by one of the two following simple rules:

(i.) *If the number is greater than unity, the characteristic is positive and numerically one less than the number of digits in its integral part.*

For let N denote a number that has n digits in its integral part; then N lies between 10^{n-1} and 10^n; that is,

$$N = 10^{(n-1)+\text{a fraction}}.$$

$\therefore \log N = (n-1) + $ a mantissa.

Thus 785 lies between 10^2 and 10^3;
hence $\log 785 = 2 + $ a mantissa.

(ii.) *If the number is less than unity, the characteristic is negative and numerically one greater than the number of ciphers immediately after the decimal point.*

For let N denote a decimal with n ciphers immediately after the decimal point; then N lies between $10^{-(n+1)}$ and 10^{-n}; that is,

$$N = 10^{-(n+1)+\text{a fraction}}.$$

$\therefore \log N = -(n+1) + $ a mantissa.

Thus 0.0078 lies between 10^{-3} and 10^{-2};
hence $\log 0.0078 = -3 + $ a mantissa.

The converse of rules (i.) and (ii.) may be stated as follows:

(i.) *If the characteristic of a common logarithm is* $+\text{n}$, *there are* $\text{n}+1$ *integral places in the corresponding number.*

(ii.) *If the characteristic is* $-\text{n}$, *there are* $\text{n}-1$ *ciphers immediately to the right of the decimal point in the number.*

Thus if $\log N = 3.6712$, then the number N will have four integral places. If $\log N = \bar{4}.9345$, then N will be less than 1, and

LOGARITHMS.

there will be three ciphers immediately to the right of the decimal point in the number N.

345. When a negative logarithm is to be divided by any number, the logarithm must be so modified in form that the negative part will be exactly divisible by the number.

Example. Given log $0.0785 = \bar{2}.8949$, find log $\sqrt[7]{0.0785}$.

$$\text{Log } \sqrt[7]{0.0785} = \tfrac{1}{7} \log 0.0785.$$
$$= \tfrac{1}{7} (\bar{2}.8949)$$
$$= \tfrac{1}{7} (\bar{7}. + 5.8949) = \bar{1}.8421.$$

Adding both -5 and $+5$ to the logarithm does not change its value and makes its negative part divisible by 7.

Exercise 100.

Given $\log 2 = 0.3010$; $\log 3 = 0.4771$; $\log 5 = 0.6990$; $\log 7 = 0.8451$; $\log 11 = 1.0414$; find the common logarithm of

1. 0.0105. $105 = 3 \times 5 \times 7$.

   $\therefore \log 105 = \log 3 + \log 5 + \log 7$
   $= 0.4771 + 0.6990 + 0.8451$
   $= 2.0212$.

   $\therefore \log 0.0105 = \bar{2}.0212$ or $8.0212 - 10$.   § 343

2. 15.	11. 8.4.	20. $42 \div 25$.	29. $\sqrt[3]{0.048}$.
3. 21.	12. 0.128.	21. $7\tfrac{1}{4}$.	30. $\sqrt[4]{0.0054}$.
4. 56.	13. 0.0035.	22. 21^5.	31. $\sqrt[6]{0.00021}$.
5. 112.	14. 0.00021.	23. $5^{\tfrac{1}{3}}$.	32. $\sqrt[5]{0.0105}$.
6. 216.	15. 0.0216.	24. $12^{\tfrac{2}{3}}$.	33. $\sqrt[7]{0.0165}$.
7. 147.	16. 0.0054.	25. $15^{\tfrac{3}{5}}$.	34. $\sqrt[11]{0.0693}$.
8. 34.3.	17. 0.0063.	26. $\sqrt[9]{98}$.	35. $(15 \div 14)^{\tfrac{1}{7}}$.
9. 37.5.	18. $77 \div 15$.	27. $\sqrt[13]{126}$.	36. $(0.21 \div 0.11)^{\tfrac{1}{5}}$.
10. 840.	19. $21 \div 16$.	28. $\sqrt{0.231}$.	37. $(0.08 \div 0.03)^{\tfrac{1}{7}}$.

LOGARITHMS.

346. Tables of Logarithms. Common logarithms have two great practical advantages: (i.) Characteristics are known by § 344, so that only mantissas are tabulated; (ii.) mantissas are determined by the sequence of digits (§ 343), so that the mantissas of integral numbers only are tabulated.

At the close of this chapter will be found a table which contains the *mantissas* of the common logarithms of all numbers from 1 to 999 correct to four decimal places.

NOTE. Tables are published which give the logarithms of all numbers from 1 to 99999 calculated to seven places of decimals; these are called 'seven-place' logarithms. For many purposes, however, the four-place or five-place logarithms are sufficiently accurate.

From a table of logarithms we may obtain (i.) the logarithm of a given number, or (ii.) the number corresponding to a given logarithm.

347. *To find the logarithm of a given number.*

(1) To find log 7.85.

By § 343 the required mantissa is the mantissa of log 785.

Look in column headed "N" for 78. Passing along this line to column headed 5, we find .8949, the required mantissa.

Adding the characteristic, we have

$$\log 7.85 = 0.8949.$$

(2) To find log 4273.2.

When the number contains more than three significant figures, use must be made of the principle that when the difference of two numbers is *small* compared with either of them, the difference of the numbers is approximately proportional to the difference of their logarithms.

By § 343 the required mantissa is that of log 427.32.

The mantissa of log 427 = .6304.
The mantissa of log 428 = .6314.

LOGARITHMS.

That is, an increase of 1 in the number causes an increase of .0010 in the mantissa; hence an increase of .32 in the number will cause an increase of .32 of .0010, or .0003, in the mantissa. Adding .0003 to the mantissa of log 427, and writing the characteristic, we have log $4273.2 = 3.6307$.

(3) To find log 0.0006049.

By § 343 the required mantissa is that of log 604.9.

The mantissa of log $604 = .7810$.
The mantissa of log $605 = .7818$.

Hence .9 of .0008, or .0007, must be added to .7810.

$\therefore$ log $0.0006049 = \overline{4}.7817$, or $6.7817 - 10$.

(4) To find log 30 or log 3, find mantissa of log 300.

348. *To find a number when its logarithm is given.*

(1) To find the number of which the logarithm is 3.8954.

Look in the table for the mantissa .8954. It is found in line 78 and in column 6; hence .8954 is the mantissa of log 786.

$\therefore 3.8954 = \log 7860;$

or 7860 is the number whose logarithm is 3.8954.

(2) To find the number of which the logarithm is 1.6290.

Look in the table for the mantissa .6290. It cannot be found; but the next less mantissa is 6284, and the next greater is 6294.

Also, $\qquad 6284 =$ mantissa of log 425,
and $\qquad 6294 =$ mantissa of log 426.

That is, an increase of .0010 in the mantissa causes an increase of 1 in the number; hence an increase of .0006 in the mantissa will cause an increase of $\frac{6}{10}$ of 1, or .6, in the number; hence .6290 is the mantissa of log 425.6; and therefore $1.6290 = \log 42.56$.

(3) To find the number of which the logarithm is $\overline{3}.8418$.

Look in the table for the mantissa 8418. It cannot be found;
but $\qquad 8414 =$ mantissa of log 694,
and $\qquad 8420 =$ mantissa of log 695.

LOGARITHMS.

That is, an increase of .0006 in the mantissa causes an increase of 1 in the number; hence an increase of .0004 in the mantissa will cause an increase of $\frac{4}{6}$ of 1, or .66, in the number; hence 8418 is the mantissa of log 694.66.

$$\therefore \overline{3}.8418 = \log 0.0069466.$$

Exercise 101.

Find, from the table, the logarithm of

1. 8.
2. 50.
3. 6.3.
4. 374.
5. 703.
6. 7.89.
7. 0.178.
8. 3.476.
9. 0.05307.
10. 78542.
11. 0.50438.
12. 0.00716.
13. 7.4803.
14. 2063.4.
15. 0.0087741.
16. 0.017423.

Find the number of which the logarithm is

17. 1.8797.
18. 7.6284−10.
19. 0.2165.
20. 2.7364.
21. 4.0095.
22. 8.1648−10.
23. 9.3178−10.
24. 1.6482.
25. 8.5209−10.
26. 3.8016.
27. 3.7425.
28. 7.1342−10.
29. 3.7045.
30. 8.7982−10.
31. 3.4793.

349. The **Cologarithm** of a number is the logarithm of its reciprocal.

That is, $\operatorname{colog} N = \log(1 \div N) = -\log N$
$$= (10 - \log N) - 10.$$

If $\log N > 0$ and < 10, colog N is written
$$(10 - \log N) - 10,$$
to avoid a negative decimal in the cologarithm.

If $\log N > 10$ and < 20, colog N is written
$$(20 - \log N) - 20,$$
for the same reason.

LOGARITHMS.

Thus $\text{colog } 0.0574 = -(\overline{2}.7589) = 1.2411$;
$\text{colog } 432 = (10 - 2.6263) - 10 = 7.3737 - 10$;
$\text{colog } 345000000000 = (20 - 11.5378) - 20 = 8.4622 - 20$.

Instead of subtracting the logarithm of a divisor, we may by § 73 add its cologarithm.

EXAMPLE. Find the value of $\dfrac{15.08 \times 0.0723}{0.0534 \times 7.238}$.

$\log 15.08 = 1.1784$
$\log 0.0723 = 8.8591 - 10$
$\text{colog } 0.0534 = 1.2725$
$\text{colog } 7.238 = 9.1404 - 10$

Add, $\log \text{(fraction)} = 0.4504 \qquad = \log 2.8213$.
Hence the fraction $= 2.8213$.

EXAMPLE 2. Find the value of $0.0543 \times 6.34 \times (-5.178)$.

$\log 0.0543 = 8.7348 - 10$
$\log 6.34 = 0.8021$
$\log 5.178 = 0.7141$

Add, $\log \text{(product)} = 0.2510 \qquad = \log 1.7824$.
Hence the product is -1.7824.

NOTE. By logarithms we obtain simply the arithmetical value of the result; its quality must be determined by the laws of quality.

EXAMPLE 3. Find the value of $\sqrt[5]{\dfrac{5.42 \times 427.^2}{3.24^4 \times 0.0231^{\frac{1}{2}}}}$.

$\log 5.42 = 0.7340 \qquad\qquad = 0.7340$
$2 \log 427. = (2.6304) \times 2 \qquad = 5.2608$
$4 \text{ colog } 3.24 = (9.4895 - 10) \times 4 = 7.9580 - 10$
$\tfrac{1}{2} \text{ colog } 0.0231 = (1.6364) \div 2 \qquad = 0.8182$

$ 5\overline{)4.7710}$
$\therefore \log \text{(root)} = 0.9542$
$\therefore \text{root} = 9$.

LOGARITHMS.

350. An Exponential Equation is one in which the unknown letter appears in an exponent; as $2^x = 5$, $x^x = 10$. Such equations are solved by the aid of logarithms.

Example 1. Solve $\quad 3^{2x} - 14 \times 3^x + 45 = 0.$ (1)

From (1), $\quad (3^x - 9)(3^x - 5) = 0.$ (2)

Equation (2) is equivalent to the two equations
$$3^x = 9 \text{ and } 3^x = 5.$$

From $\quad 3^x = 9, x = 2$; and from $3^x = 5$,
$$x \log 3 = \log 5.$$
$$\therefore x = \frac{\log 5}{\log 3} = \frac{0.6990}{0.4771} = 1.4649.$$

Hence the solutions of (1) are 2 and 1.4649.

Exercise 102.

Find by logarithms the value of

1. $742.8 \times 0.02374.$
2. $0.3527 \times 0.00572.$
3. $78.42 \times 0.000437.$
4. $5234 \times (-0.03671).$
5. $3.246 \times (-0.0746).$
6. $-4.278 \times (-0.357).$
7. $4743 \div 327.4.$
8. $9.345 \div (-0.0765).$
9. $\dfrac{2.476 \times (-0.742)}{73.81 \times (-0.00121)}.$
10. $\dfrac{321 \times (-48.1) \times (357)}{421 \times (-741) \times (4.21)}.$

11. $5^{\frac{3}{4}}.$
12. $0.021^{\frac{2}{3}}.$
13. $0.532^8.$
14. $(\frac{17}{14})^9.$
15. $714.2^{\frac{2}{3}}.$
16. $(\frac{471}{801})^7.$
17. $(\frac{15}{23})^{2.7}.$
18. $(3\frac{2}{5})^{1.27}.$
19. $4.71^{3.205}.$

20. $\sqrt{\dfrac{0.035^3 \times 54.2 \times 785^{\frac{1}{4}} \times 0.0742}{4.72^{\frac{1}{4}} \times 7.14^{\frac{1}{5}} \times 8.47^{\frac{1}{6}}}}.$

LOGARITHMS.

21. $\sqrt[3]{\dfrac{0.0427^2 \times 5.27 \times 0.875^4}{7.421^{\frac{1}{4}} \times \sqrt{1.74} \times \sqrt{0.00215}}}$.

22. $\sqrt[5]{\dfrac{0.714^{\frac{1}{2}} \times 0.1371^{\frac{1}{3}} \times 0.0718^{\frac{1}{4}}}{0.524^2 \times 0.742^{\frac{1}{3}} \times 0.0527^{\frac{1}{2}}}}$.

Solve

23. $31^x = 23$. 25. $5^x = 800$. 27. $5^{x-3} = 8^{2x+1}$.
24. $0.3^x = 0.8$. 26. $12^x = 3528$. 28. $a^{2x} b^{3x} = c^5$.
29. $2^{3x} 5^{2x-1} = 4^{5x} 3^{x+1}$. 30. $4^{2x} + 56 = 15 \times 4^x$.

COMPOUND INTEREST AND ANNUITIES.

351. *To find the compound interest, $ I, and amount, $ M, of a given principal, $ P, in n years, $ r being the interest on $ 1 for 1 year.*

Let $ R = $ the amount of $ 1 in 1 year; then $R = 1 + r$, and the amount of $ P at the end of the first year is $ PR; and since this is the principal for the second year, the amount at the end of the second year is $ PR \times R$, or $ PR^2$. For like reason the amount at the end of the third year is $ PR^3$, and so on; hence the amount in n years is $ PR^n$; that is,

$$M = PR^n, \text{ or } P(1+r)^n. \qquad (1)$$

Hence $\qquad I = P(R^n - 1). \qquad (2)$

If the interest is payable semi-annually, the amount of $ P in $\frac{1}{2}$ a year will be $ P(1 + \frac{1}{2}r)$; hence, as n years equals $2n$ half-years,

$$M = P(1 + \tfrac{1}{2}r)^{2n}. \qquad (3)$$

Similarly, if the interest is payable quarterly,

$$M = P(1 + \tfrac{1}{4}r)^{4n}. \qquad (4)$$

LOGARITHMS.

Example. Find the time in which a sum of money will double itself at ten per cent compound interest, interest to be "converted into principal" semi-annually.

Here $1 + \frac{1}{2}r = 1.05$. Let $P = 1$; then $M = 2$.

Substituting these values in (3), we obtain

$$2 = (1.05)^{2n}.$$

$$\therefore \log 2 = 2n \cdot \log 1.05.$$

$$\therefore n = \frac{\log 2}{2 \log 1.05} = \frac{0.3010}{0.0424} = 7.1 \text{ years. } Ans.$$

352. Present Value and Discount. Let $\$P$ denote the present value of the sum $\$M$ due in n years, at the rate $\$r$; then evidently in n years at the rate $\$r$, $\$P$ will amount to $\$M$; hence

$$M = PR^n, \text{ or } P = MR^{-n}.$$

Let $\$D$ be the discount; then

$$D = M - P = M(1 - R^{-n}).$$

353. An **Annuity** is a fixed sum of money that is payable once a year, or at more frequent regular intervals, under certain stated conditions. An *Annuity Certain* is one payable for a fixed number of years. A *Life Annuity* is one payable during the lifetime of a person. A *Perpetual Annuity*, or *Perpetuity*, is one that is to continue forever, as, for instance, the rent of a freehold estate.

354. *To find the amount of an annuity left unpaid for a given number of years, allowing compound interest.*

Let $\$A$ be the annuity, n the number of years, $\$R$ the amount of one dollar in one year, $\$M$ the required amount. Then evidently the number of dollars due at the end of the

LOGARITHMS.

First year $= A$;
Second year $= AR + A$;
Third year $= AR^2 + AR + A$;
nth year $= AR^{n-1} + AR^{n-2} + \cdots + AR + A$
$$= \frac{A(R^n - 1)}{R - 1}.$$

That is, $\quad M = \dfrac{A}{r}(R^n - 1).\quad$ (1)

EXAMPLE 1. Find the amount of an annuity of $100 in 20 years, allowing compound interest at $4\frac{1}{2}$ per cent.

$$M = \frac{A}{r}(R^n - 1) = \frac{100\,(1.045^{20} - 1)}{0.045}.$$

By logarithms $\quad 1.045^{20} = 2.4117.$

$$\therefore M = \frac{141.17}{0.045} = 3137.11.$$

Hence the amount of the annuity is $3137.11.

EXAMPLE 2. What sum must be set aside annually that it may amount to $50,000 in 10 years at 6 per cent compound interest?

From (1), $\quad A = \dfrac{Mr}{R^n - 1} = \dfrac{50{,}000 \times 0.06}{1.06^{10} - 1} = 3793.37.$

Hence the required sum is $3793.37.

355. *To find the present value of an annuity of* $A *payable at the end of each of* n *successive years.*

Let $P denote the present value; then the amount of $P in n years will equal the amount of the annuity in the same time; that is,

$$PR^n = A(R^n - 1)r^{-1}. \quad (1)$$
$$\therefore P = A(1 - R^{-n})r^{-1}. \quad (2)$$

If the annuity be *perpetual*, then $n = \infty$, $R^{-n} \doteq 0$, and (2) becomes
$$P = Ar^{-1}.$$

LOGARITHMS.

Exercise 103.

1. Write out the logarithmic equations for finding each of the four numbers M, R, P, n.

2. In what time, at 5 per cent compound interest, will $100 amount to $1000?

3. Find the time in which a sum will double itself at 4 per cent compound interest.

4. Find in how many years $1000 will become $2500 at 10 per cent compound interest.

5. Find the present value of $10,000 due 8 years hence at 5 per cent compound interest.

6. Find the amount of $1 at 5 per cent compound interest in a century.

7. Show that money will increase more than seventeen-thousand-fold in a century at 10 per cent compound interest.

8. If A leaves B $1000 a year to accumulate for 3 years at 4 per cent compound interest, find what amount B should receive.

9. Find the present value of the legacy in Example 8.

10. Find the present value, at 5 per cent, of an estate of $1000 a year to be entered on immediately.

11. A freehold estate worth $120 a year is sold for $4000; find the rate of interest.

12. A man has a capital of $20,000, for which he receives interest at 5 per cent; if he spends $1800 every year, show that he will be ruined before the end of the 17th year.

LOGARITHMS.

N	0	1	2	3	4	5	6	7	8	9
10	0000	0043	0086	0128	0170	0212	0253	0294	0334	0374
11	0414	0453	0492	0531	0569	0607	0645	0682	0719	0755
12	0792	0828	0864	0899	0934	0969	1004	1038	1072	1106
13	1139	1173	1206	1239	1271	1303	1335	1367	1399	1430
14	1461	1492	1523	1553	1584	1614	1644	1673	1703	1732
15	1761	1790	1818	1847	1875	1903	1931	1959	1987	2014
16	2041	2068	2095	2122	2148	2175	2201	2227	2253	2279
17	2304	2330	2355	2380	2405	2430	2455	2480	2504	2529
18	2553	2577	2601	2625	2648	2672	2695	2718	2742	2765
19	2788	2810	2833	2856	2878	2900	2923	2945	2967	2989
20	3010	3032	3054	3075	3096	3118	3139	3160	3181	3201
21	3222	3243	3263	3284	3304	3324	3345	3365	3385	3404
22	3424	3444	3464	3483	3502	3522	3541	3560	3579	3598
23	3617	3636	3655	3674	3692	3711	3729	3747	3766	3784
24	3802	3820	3838	3856	3874	3892	3909	3927	3945	3962
25	3979	3997	4014	4031	4048	4065	4082	4099	4116	4133
26	4150	4166	4183	4200	4216	4232	4249	4265	4281	4298
27	4314	4330	4346	4362	4378	4393	4409	4425	4440	4456
28	4472	4487	4502	4518	4533	4548	4564	4579	4594	4609
29	4624	4639	4654	4669	4683	4698	4713	4728	4742	4757
30	4771	4786	4800	4814	4829	4843	4857	4871	4886	4900
31	4914	4928	4942	4955	4969	4983	4997	5011	5024	5038
32	5051	5065	5079	5092	5105	5119	5132	5145	5159	5172
33	5185	5198	5211	5224	5237	5250	5263	5276	5289	5302
34	5315	5328	5340	5353	5366	5378	5391	5403	5416	5428
35	5441	5453	5465	5478	5490	5502	5514	5527	5539	5551
36	5563	5575	5587	5599	5611	5623	5635	5647	5658	5670
37	5682	5694	5705	5717	5729	5740	5752	5763	5775	5786
38	5798	5809	5821	5832	5843	5855	5866	5877	5888	5899
39	5911	5922	5933	5944	5955	5966	5977	5988	5999	6010
40	6021	6031	6042	6053	6064	6075	6085	6096	6107	6117
41	6128	6138	6149	6160	6170	6180	6191	6201	6212	6222
42	6232	6243	6253	6263	6274	6284	6294	6304	6314	6325
43	6335	6345	6355	6365	6375	6385	6395	6405	6415	6425
44	6435	6444	6454	6464	6474	6484	6493	6503	6513	6522
45	6532	6542	6551	6561	6571	6580	6590	6599	6609	6618
46	6628	6637	6646	6656	6665	6675	6684	6693	6702	6712
47	6721	6730	6739	6749	6758	6767	6776	6785	6794	6803
48	6812	6821	6830	6839	6848	7857	6866	6875	6884	6893
49	6902	6911	6920	6928	6937	6946	6955	6964	6972	6981
50	6990	6998	7007	7016	7024	7033	7042	7050	7059	7067
51	7076	7084	7093	7101	7110	7118	7126	7135	7143	7152
52	7160	7168	7177	7185	7193	7202	7210	7218	7226	7235
53	7243	7251	7259	7267	7275	7284	7292	7300	7308	7316
54	7324	7332	7340	7348	7356	7364	7372	7380	7388	7396

LOGARITHMS.

N	0	1	2	3	4	5	6	7	8	9
55	7404	7412	7419	7427	7435	7443	7451	7459	7466	7474
56	7482	7490	7497	7505	7513	7520	7528	7536	7543	7551
57	7559	7566	7574	7582	7589	7597	7604	7612	7619	7627
58	7634	7642	7649	7657	7664	7672	7679	7686	7694	7701
59	7709	7716	7723	7731	7738	7745	7752	7760	7767	7774
60	7782	7789	7796	7803	7810	7818	7825	7832	7839	7846
61	7853	7860	7868	7875	7882	7889	7896	7903	7910	7917
62	7924	7931	7938	7945	7952	7959	7966	7973	7980	7987
63	7993	8000	8007	8014	8021	8028	8035	8041	8048	8055
64	8062	8069	8075	8082	8089	8096	8102	8109	8116	8122
65	8129	8136	8142	8149	8156	8162	8169	8176	8182	8189
66	8195	8202	8209	8215	8222	8228	8235	8241	8248	8254
67	8261	8267	8274	8280	8287	8293	8299	8306	8312	8319
68	8325	8331	8388	8344	8351	8357	8363	8370	8376	8382
69	8388	8395	8401	8407	8414	8420	8426	8432	8439	8445
70	8451	8457	8463	8470	8476	8482	8488	8494	8500	8506
71	8513	8519	8525	8531	8537	8543	8549	8555	8561	8567
72	8573	8579	8585	8591	8597	8603	8609	8615	8621	8627
73	8633	8639	8645	8651	8657	8663	8669	8675	8681	8686
74	8692	8698	8704	8710	8716	8722	8727	8733	8739	8745
75	8751	8756	8762	8768	8774	8779	8785	8791	8797	8802
76	8808	8814	8820	8825	8831	8837	8842	8848	8854	8859
77	8865	8871	8876	8882	8887	8893	8899	8904	8910	8915
78	8921	8927	8932	8938	8943	8949	8954	8960	8965	8971
79	8976	8982	8987	8993	8998	9004	9009	9015	9020	9025
80	9031	9036	9042	9047	9053	9058	9063	9069	9074	9079
81	9085	9090	9096	9101	9106	9112	9117	9122	9128	9133
82	9138	9143	9149	9154	9159	9165	9170	9175	9180	9186
83	9191	9196	9201	9206	9212	9217	9222	9227	9232	9238
84	9243	9248	9253	9258	9263	9269	9274	9279	9284	9289
85	9294	9299	9304	9309	9315	9320	9325	9330	9335	9340
86	9345	9350	9355	9360	9365	9370	9375	9380	9385	9390
87	9395	9400	9405	9410	9415	9420	9425	9430	9435	9440
88	9445	9450	9455	9460	9465	9469	9474	9479	9484	9489
89	9494	9499	9504	9509	9513	9518	9523	9528	9533	9538
90	9542	9547	9552	9557	9562	9566	9571	9576	9581	9586
91	9590	9595	9600	9605	9609	9614	9619	9624	9628	9633
92	9638	9643	9647	9652	9657	9661	9666	9671	9675	9680
93	9685	9689	9694	9699	9703	9708	9713	9717	9722	9727
94	9731	9736	9741	9745	9750	9754	9759	9763	9768	9773
95	9777	9782	9786	9791	9795	9800	9805	9809	9814	9818
96	9823	9827	9832	9836	9841	9845	9850	9854	9859	9863
97	9868	9872	9877	9881	9886	9890	9894	9899	9903	9908
98	9912	9917	9921	9926	9930	9934	9939	9943	9948	9952
99	9956	9961	9965	9969	9974	9978	9983	9987	9991	9996

LOGA

195. Logarithms are special ex~~~~~~~.

Every positive number can be expressed exactly or approximately as a power of 10; as $100 = 10^2$.

The exponent required is called the **logarithm of the number to the base 10.**

Thus 2 is the logarithm of 100 to the base 10.
It is written briefly thus: $2 = \log_{10} 100$, or $2 = \log 100$.

196. How some logarithms can be obtained.

$10^0 = 1$, or $0 = \log 1$. $10^1 = 10$, or $1 = \log 10$.
$10^{.5} = \sqrt{10} = 3.162$; or $.50 = \log 3.162$.
$10^{.25} = (10^{.5})^{\frac{1}{2}} = \sqrt{10^{.5}} = \sqrt{3.162} = 1.778$; or $.25 = \log 1.778$.

By similar computation, the table below can be obtained.

197. How logarithms are used.

Example 1. Find 3.1623×17.782.

$10^{.00}$ =	1.0000
$10^{.25}$ =	1.7782
$10^{.50}$ =	3.1623
$10^{.75}$ =	5.6234
$10^{1.00}$ =	10.0000
$10^{1.25}$ =	17.7820
$10^{1.50}$ =	31.6230
$10^{1.75}$ =	56.2340
$10^{2.00}$ =	100.0000

Solution. 1. 3.1623×17.782
2. $= 10^{.50} \times 10^{1.25}$
3. $= 10^{1.75}$
4. $= 56.234$ (in the table)
5. $\therefore 3.1623 \times 17.782 = 56.234$.
The solution is approximately correct.

Example 2. Find $(5.6234)^2 \times 31.623 \div 17.782$.

Solution. 1. $(5.6234)^2 \times 31.623 \div 17.782$
2. $= (10^{.75})^2 \times 10^{1.50} \div 10^{1.25}$
3. $= 10^{1.50+1.50-1.25} = 10^{1.75}$
4. $\therefore (5.6234)^2 \times 31.623 \div 17.782 = 56.234$

This solution also may be checked by ordinary computation by *one who has a lot of time and ambition.*

LOGARITHMS

198. Logarithms of numbers to the base 10 are called **common logarithms** or simply logarithms.

199. Graph of the functional relation $y = \log x$.

Besides the values of the table on page 184, observe:
$$10^{-.25} = 10^{.75-1.00} = 10^{.75} \div 10^1 = 5.6234 \div 10 = .562.$$
Similarly $10^{-.50} = 10^{.5} \div 10^1 = 3.1623 \div 10 = .3162.$

Express the values of x correct to tenths, and the values of y correct to hundredths.

When $x =$	.3	.6	1.0	1.8	3.2	5.6	10
then $y =$	$-.50$	$-.25$	0.00	.25	.50	.75	1.00

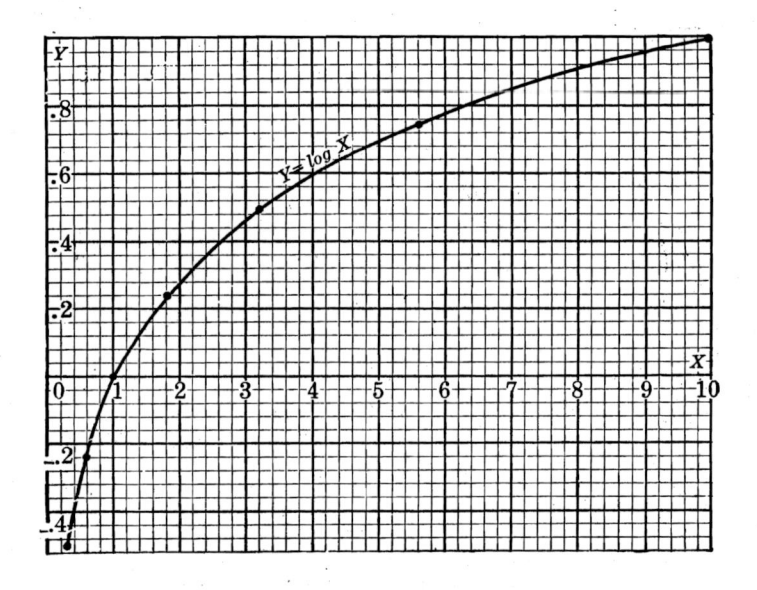

There are no values of y for negative values of x.

y is negative when x lies between 0 and 1.

y increases as x increases.

ALGEBRA

200. Most numbers are not exact powers of 10.

Thus, from the graph on page 185, log 6 = about .78
Correct to four decimal places log 6 = .7782. Similarly the log 60 is 1.7782.

The integral part of a logarithm is called the **characteristic** and the decimal part the **mantissa**.

Thus, the characteristic of log 60 is 1, and the mantissa is .7782.

201. Finding the characteristic of the logarithm of a number greater than 1.

It is known that $3.53 = 10^{.5478}$ $\therefore$ log 3.53 = .5478
$35.3 = 10 \times 3.53 = 10 \times 10^{.5478} = 10^{1.5478}$ $\therefore$ log 35.3 = 1.5478
$353 = 10 \times 35.3 = 10 \times 10^{1.5478} = 10^{2.5478}$ $\therefore$ log 353 = 2.5478

In this last line, 353 has three figures to the left of the decimal point; its logarithm has characteristic 2, which is 1 less than 3.

Rule. *The characteristic of the common logarithm of a number greater than 1 is one less than the number of significant figures to the left of the decimal point.*

Thus, the characteristic of log 357.83 is 2; of log 70390 is 4.

202. Finding the characteristic of the logarithm of a number less than 1.

$.353 = \dfrac{3.53}{10} = \dfrac{10^{.5478}}{10} = 10^{.5478-1}$ $\therefore$ log .353 = .5478 − 1

$.0353 = \dfrac{.353}{10} = \dfrac{10^{.5478-1}}{10} = 10^{.5478-2}$ $\therefore$ log .0353 = .5478 − 2

$.00353 = \dfrac{10^{.5478-2}}{10} = 10^{.5478-3}$ $\therefore$ log .00353 = .5478 − 3

Observing you will find that the following rule is correct.

Rule. *The characteristic of the common logarithm of a (positive) number less than 1 is negative; numerically it is one more than the number of zeros between the decimal point and the first significant figure.*

LOGARITHMS

203. The method of writing a negative characteristic.

In § 202 log .353 = .5478 − 1. Actually, therefore, log .353 is − .4522, a negative number. However, the positive mantissa and the negative characteristics are retained, as follows.

.5478 − 1 is written: 9.5478 − 10. Numerically the two expressions have equal value. Note that 9 − 10 = − 1.

In general, *decide upon the characteristic by the rule in* § 202; then, if it is − 1, write it 9 − 10; if − 2, write it 8 − 10; etc.; *and then omit the* − 10, *usually*.

Thus, log .02 is .3010 − 2, or 8.3010 − 10, or 8.3010.

NOTE. The negative characteristic is often written thus: log .02 = $\bar{2}$.3010; again, log .353 = $\bar{1}$.5478. The minus sign is written over the characteristic to indicate that it alone is negative, the mantissa being positive. Your teacher will decide which of these two methods you will use.

EXERCISE 112

Find the characteristic of the logarithm of:

1. 59	**5.** 72,860	**9.** 5.08	**13.** 984.2
2. 540	**6.** 11.2	**10.** 3002	**14.** 87,600
3. 4000	**7.** 367.2	**11.** 21.67	**15.** 2.193
4. 8	**8.** 50900	**12.** 100.5	**16.** 1,000,000

Tell the number of significant figures preceding the decimal point when the characteristic of the logarithm is:

17. 5   **18.** 3   **19.** 0   **20.** 1   **21.** 4   **22.** 2

Write in two ways the characteristic of the logarithm of:

23. .5	**25.** .07	**27.** .6432	**29.** .1007
24. .004	**26.** .01003	**28.** .04216	**30.** .00008

Tell the number of zeros preceding the first significant figure when the characteristic of the logarithm is:

31. − 2   **32.** − 4   **33.** − 1   **34.** − 5   **35.** − 3

ALGEBRA

204. The mantissa of the logarithm of a number. From the illustrations in § 201 to § 203, it is clear that *the common logarithms of all numbers having the same significant figures have the same mantissas.* These mantissas are given in a table of logarithms such as appears on pages 254 and 255.

205. Finding the logarithm of a three digit number.

Example 1. Find the logarithm of 16.8.

Solution. 1. In the column headed "No." (page 254) find 16. On the horizontal line opposite 16, pass over to the column headed by the figure 8. The mantissa 2253 found there is the required mantissa.

2.   The characteristic is 1, by the rule in § 201.

3.   ∴ log 16.8 is 1.2253.

Rule. **To find the logarithm of a number of three figures:**

1. *Look in the column headed "No." (pages 254–255) for the first two figures of the given number. The mantissa will be found on the horizontal line opposite these two figures and in the column headed by the third figure of the given number.*

2. *Prefix the characteristic according to § 201 and § 202.*

Example 2. Find log .304.

Solution. 1. Opposite 30 in the column headed by 4 is the mantissa .4829. The characteristic is -1 or $9 - 10$. (§ 202 and § 203.)

2.   ∴ **log** .304 = 9.4829 − 10, = 9.4829.

NOTE. The logarithm of a number of one or two digits may be found by using the column headed 0. Thus the mantissa of log 8.3 is the same as the mantissa of log 8.30; of log 9, the same as of log 900.

EXERCISE 113

Find the logarithm of:

1. 365	6. 64	11. .841	16. .000834
2. 571	7. 9	12. .0628	17. .07
3. 847	8. 5.2	13. .00175	18. 3.14
4. 902	9. 43.6	14. 7680	19. 40.8
5. 200	10. 720	15. 25900	20. .16

LOGARITHMS

206. The logarithm of a number of more than three digits.

Example 1. Find log 327.5.

Solution. 1. From the table on page 254.
log 327   = 2.5145
log 327.5 = ?
log 328   = 2.5159
} Difference = .0014.

2. Since 327.5 is between 327 and 328, its logarithm must be between their logarithms. An increase of one unit in the number (from 327 to 328) produces an increase of .0014 in the mantissa. It is *assumed* therefore that an increase of .5 in the number (from 327 to 327.5) produces an increase of .5 of .0014, or of .0007 in the mantissa.

3. ∴ log 327.5 = 2.5145 + .0007, or 2.5152.

.0014 is called the *tabular difference*. The zeros are usually omitted.

Example 2. Find log 34.67.

Solution. 1. Mantissa of log 346 = 5391
Mantissa of log 347 = 5403
2. Tabular difference = 12.   .7 × 12 = 8.4 or 8
3. ∴ mantissa for log 3467 = 5391 + 8, or 5399
4. ∴ log 34.67 = 1.5399.

Rule. 1. *Find the mantissa for the first three figures, and the tabular difference for that mantissa.*

2. *Multiply the tabular difference by the remaining figures of the given number, preceded by a decimal point.*

3. *Add the result of Step 2 to the mantissa obtained in Step 1, writing the sum correct to four places.*

4. *Prefix the proper characteristics.* (See Rules, page 186.)

EXERCISE 114

Find the logarithm of:

1. 342.5
2. 252.1
3. 865.2
4. 764.4
5. 438.3
6. 501.6
7. 28.25
8. 1.158
9. 7.631
10. .5842
11. .04873
12. 328.2
13. 453.3
14. 86.43
15. 3.728
16. 1.067
17. 5.243
18. 3.142
19. 632.4
20. 82.56

ALGEBRA

207. Finding the number when its logarithm is given.

NOTE. Some teachers call this finding the *antilogarithm*.

Example 1. Find the number whose logarithm is 1.6571.

Solution. 1. Find the mantissa 6571 in the table on pages 254–255.

2. In the column headed "No." on the line with 6571 is 45. At the head of the column containing 6571 is 4. Hence the number sought has the figures 454.

3. The characteristic being 1, the number must have two figures to the left of the decimal point. $\therefore$ the number is 45.4.

Example 2. Find the number whose logarithm is 1.3934.

Solution. 1. The mantissa 3934 *does not appear* in the table.
The next less mantissa is 3927, and the next greater is 3945.
That is:
$$\left. \begin{array}{l} \text{mantissa of log } 247 = 3927 \\ \text{mantissa of log } x = 3934 \\ \text{mantissa of log } 248 = 3945 \end{array} \right\} \left. \begin{array}{l} \text{Difference} \\ = 7. \end{array} \right\} \left. \begin{array}{l} \text{Tabular} \\ \text{difference} \\ = 18. \end{array} \right.$$

2. The increase of 18 in the mantissa produces an increase of 1 in the number. We *assume* that the increase of 7 produces an increase of $\frac{7}{18}$ or about .4 in the number.

3. Hence the number has the figures 247.4.

4. Since the characteristic is 1, the number is 24.74.

Example 3. Find the antilogarithm of 9.3940.

Solution. 1. Again x is between 247 and 248.

2. The difference = 13. Tabular difference = 18. $\frac{13}{18} = .7$

3. $\therefore$ the number has the figures 247.7.

4. Since the characteristic is 9, the number is .2477.

EXERCISE 115

Find the number whose logarithm is

1. 2.7408	6. 9.5969(−10)	11. 2.2930	16. 8.9194
2. 1.6678	7. 8.3429(−10)	12. 1.9665	17. 7.7978
3. .4188	8. 3.8497	13. 3.8598	18. 3.2306
4. 3.8983	9. 7.3288(−10)	14. 9.7606	19. 1.8817
5. 2.9417	10. .1195	15. 4.3346	20. 2.9986

LOGARITHMS

The Operations with Logarithms

208. The logarithm of a product *equals the sum of the logarithms of the factors.*

That is: $\quad\quad \text{Log } MN = \log M + \log N$

Proof 1. $\quad\quad$ Let $\quad M = 10^x$; or $x = \log M$
$\quad\quad\quad\quad\quad\quad$ and let $N = 10^y$; or $y = \log N$

2. $\quad\quad\quad\quad\quad\quad MN = 10^{x+y}$
3. $\therefore \log MN = x + y$; or $\log MN = \log M + \log N$.

Example. Find the value of $7.208 \times .0631$.

Solution. 1. Let $v = 7.208 \times .0631$. $\quad\quad |\ \log 7.208 = .8578$
2. $\therefore \log v = \log 7.208 + \log .0631 \quad |\ \log .0631 = \underline{8.8000 - 10}$
3. $\therefore \log v = 9.6578 \quad\quad\quad\quad\quad\quad\quad\quad\quad\quad\quad\quad 9.6578 - 10$
4. $\therefore\quad v = .4548$

By the suggestion on page 68, it may be preferable to write this as .455, since there are only three significant figures in .0631.

At any rate, do not keep more than four significant figures.

EXERCISE 116

Find the following products by logarithms:

1. 42.5×26.8
2. 3.89×72.6
3. $535 \times .621$
4. $.342 \times 2.15$
5. $.654 \times .368$
6. 239.5×38.4
7. 871.2×45
8. 1.414×360
9. $8.42 \times .793$
10. $6.282 \times .778$
11. 7.026×8059
12. 432.4×1.658
13. 93.62×768.7
14. $84.75 \times .036$
15. $76.85 \times .0043$
16. $C = 2\pi r$. Find C when $\pi = 3.142$ and $r = 13$.
17. $S = \pi r h$. Find S, when $\pi = 3.142$; $r = 11$; $h = 16$.
18. $I = prt$. Find I, when $p = \$2750$; $r = .06$; $t = 4.5$.
19. $\text{Log } x = .4771$. Find $\log x^2$.
20. $\text{Log } y = .3010$. Find $\log 10\ y$.
21. Find $84.75 \times .368 \times 3.14$ by logarithms.
22. Find by logarithms the value of $(73.84)^2$.

ALGEBRA

209. The logarithm of the quotient *of two numbers is the logarithm of the dividend minus the logarithm of the divisor.*

That is: **Log** $(M \div N) = \log M - \log N$.

Proof 1. Let $M = 10^x$, or $x = \log M$
and let $N = 10^y$, or $y = \log N$
2. $\therefore M \div N = 10^{x-y}$
3. $\therefore \log(M \div N) = x - y$; or $\log(M \div N) = \log M - \log N$.

Example 1. Find the value of $215 \div 7.25$.

Solution. 1. Let $v = 215 \div 7.25$
2. $\therefore \log v = \log 215 - \log 7.25$
3. $\therefore \log v = 1.4721$
4. $\therefore \quad v = 29.65$
or $v = 29.6$. (See p. 68.)

$\log 215 =$	2.3324
$\log 7.25 =$	$.8603$
Subtract	1.4721

Find v by the rule on **page 190**.

Example 2. Find the value of $.192 \div .216$.

Solution. 1. Log $v = \log .192 - \log .216$
2. $\quad \log v = 9.9488$
3. $\quad v = .888$

$\log .192 =$	$19.2833 - 20$
$\log .216 =$	$9.3345 - 10$
	$9.9488 - 10$

NOTE. $\log .192 = 9.2833 - 10$. If we subtract 9.3345 from this, we get a negative result. Since the mantissas in the table are all positive numbers, we write $9.2833 - 10$ as $19.2833 - 20$.

210. Optional method for division. Observe that

$$M \div N = M \times \left(\frac{1}{N}\right); \text{ so } \log M \div N = \log M + \log\left(\frac{1}{N}\right)$$

The $\log \left(\frac{1}{N}\right)$ is called the **cologarithm** of N.

Now $\log \left(\frac{1}{N}\right) = \log 1 - \log N = 0 - \log N = 10 - \log N - 10.$

The advantage is that *cologarithms are added*.

Thus: $\log \dfrac{75}{26 \times 1.8} = \log 75 \quad = 1.8751$
$\phantom{\log \dfrac{75}{26 \times 1.8}} + \text{colog } 26 = 8.5850 - 10$
$\phantom{\log \dfrac{75}{26 \times 1.8}} + \text{colog } 1.8 = 9.7447 - 10$
$\therefore \log \dfrac{75}{26 \times 1.8} \phantom{+ \text{colog } 1.8} = 20.2048 - 20.$

Colog $26 =$
$10 - 1.4150 - 10$
$= 8.5850 - 10$
Colog $1.8 =$
$10 - .2553 - 10$
$= 9.7447 - 10.$

LOGARITHMS

211. **The logarithm of a power (or a root) of a number is the logarithm of the number multiplied by its exponent;** that is
$$\log M^p = p \log M.$$

Proof 1.   Let $M = 10^x$
2. $\therefore \quad M^p = (10^x)^p = 10^{xp}$
3. $\therefore \log M^p = xp$
4. $\therefore \log M^p = p \log M.$

Example 1. Find by logarithms 1.04^{10}.
Solution. 1. $\log 1.04^{10} = 10 \log 1.04 = 10 \times .0170 = .1700$
2. $\therefore 1.04^{10} = 1.479.$

Example 2. Find by logarithms $\sqrt[4]{.0359}$.
Solution. 1. $\log \sqrt[4]{.0359} = \frac{1}{4} \log .0359,$ or $\frac{1}{4}(8.5551 - 10)$
2. $\therefore \log \sqrt[4]{.0359} = \frac{1}{4}(38.5551 - 40)$ (See note below.)
3. $\therefore \log \sqrt[4]{.0359} = 9.6387 - 10$
4. $\therefore \quad\quad \sqrt[4]{.0359} = .4353.$

NOTE. To divide a negative logarithm, write it in such form that the negative part of the characteristic may be divided exactly by the divisor, and give -10 as quotient.

EXERCISE 117

Find by logarithms the value of:

1. $335 \div 56$
2. $483 \div 71$
3. $230.4 \div 125$
4. $739.8 \div 1.73$
5. $3305 \div 1.414$
6. $8.964 \div 45.25$
7. $\dfrac{4.16 \times 32}{485}$
8. $\dfrac{35.2 \times 1.52}{53.87}$
9. $\dfrac{43.57 \times .069}{3.14}$
10. $\dfrac{14.07 \times 347}{18}$
11. $\dfrac{3.25 \times .0063}{.007}$
12. $\dfrac{527.8 \times .069}{2.449}$
13. 323^2
14. 4.025^2
15. $\sqrt{418.5}$
16. $\sqrt[3]{784}$
17. $\sqrt[4]{92.04}$
18. 8.975^2
19. $\frac{4}{3} \times 3.142 \times 6.5^3$
20. 3.142×19^2

ALGEBRA

EXERCISE 118

Use logarithms to find products, quotients, powers, etc.

1. By the formula $I = prt$.
 (a) Find I, if $p = \$4250$; $r = 4\%$; $t = 4$ yr.
 (b) Find p if $I = \$375$; $r = 5\%$; $t = 3.5$ yr.
 (c) Find t if $I = \$840$; $r = 6\%$; and $p = \$2200$.
 (d) Find r if $I = \$750$; $p = \$3750$; and $t = 4$.

2. $z = 2\pi rh$. ($\pi = 3.142$)
 (a) Find z when $r = 13$; and $h = 7.5$.
 (b) Find r when $h = 11.2$; and $z = 628$.
 (c) Find h when $z = 964$ and $r = 6.5$.

3. $t = \pi\sqrt{\dfrac{l}{g}}$. ($\pi = 3.142$; $g = 32.16$)
 (a) Find t when $l = 30$.
 (b) Find l when $t = 2$.

4. $V = \pi r^2 h$.
 (a) Find V when $r = 12.4$ and $h = 30.3$.
 (b) Find h when $V = 20250$ and $r = 15$.
 (c) Find r if $h = 575$, and $V = 8550$.

5. $V = \frac{1}{3}\pi r^2 h$.
 (a) Find V when $r = 6$ and $h = 13$.
 (b) Find h when $r = 7.5$ and $V = 630$.
 (c) Find r when $V = 725$ and $h = 12$.

6. $S = \frac{1}{2} gt^2$. ($g = 32.16$)
 (a) Find S when $t = 4.5$.
 (b) Find t when $S = 375$.

7. $A = p\left(1 + \dfrac{r}{100}\right)^n$ is the formula for the amount to which p dollars will accumulate at $r\%$, compounded annually for n years. (a) Find A if $p = \$450$; $r = 4$; $n = 8$.
 (b) Find p if $A = 20000$; $r = 4$; $n = 10$.

LOGARITHMS

EXERCISE 119. CUMULATIVE REVIEW

1. Find the prime factors of each of the following:
   (a) $10a - 11a^2 - 6a^3$ (d) $x^5 - \frac{1}{32}y^5$
   (b)* $x^3 + 2x^2 - 5x - 6$ (e) $225x^2 + 30mn - 9n^2 - 25m^2$
   (c) $9x^{4a} - 12x^{2a} + 4$ (f) $9(a+b)^2 - 25$

2. Rationalize the denominator of $\dfrac{\sqrt{15}}{\sqrt{5} - \sqrt{3}}$

3. Solve and check the equation
$$3\sqrt{x+1} + x = 9$$

4. Find the roots of $.2x^2 - .32x = .4$ to the nearest hundredth.

5. Solve the system $\begin{cases} a^2 + ab + b^2 = 19 \\ a^2 - ab + b^2 = 7 \end{cases}$
   Group your results, and check.

6. (a) Multiply $x^{\frac{2}{3}} - 2x^{\frac{1}{3}} + 3$ by $x^{\frac{1}{3}} - 1$.
   (b) Express the result without any fractional exponents.

7. (a) Without solving the equation, determine the nature of the roots of the equation $4x^2 - 7x + 9 = 0$.
   (b) Form the equation whose roots are $-\frac{1}{2}$ and $\frac{2}{3}$.
   (c) For what value of k will the roots of $2x^2 - 5x + k = 0$ be equal?

8. By the use of logarithms, find the value of
$$\frac{(2.53) \times \sqrt{15.2}}{.8514}$$

9. A boat can go 15 miles upstream in the time that it requires to go 25 miles downstream. How does the rate of the boat in still water compare with the rate of the stream?

10. Solve the system $\begin{cases} x^2 = y \\ x + 2y = 6 \end{cases}$ graphically.

11. Determine graphically the roots of $3x^2 - 2x = 8$.

12. Determine graphically the roots of $3x^2 - 2x = 15$.

LOGARITHMS

170. Fractional Power. If we are asked to find the value of 2^3, we know that it stands for $2 \times 2 \times 2$, or 8. But symbols such as $3^{\frac{1}{2}}$ or $5^{\frac{2}{3}}$, in which the powers are fractions, are meaningless, not having yet been defined.

It is easy to find what such symbols ought to mean if we remember that $x^m x^n = x^{m+n}$, m and n being whole numbers.

1. Find the meaning of $3^{\frac{1}{2}}$.

Let $\qquad y = 3^{\frac{1}{2}}$.
Then $\qquad y \times y = 3^{\frac{1}{2}} \times 3^{\frac{1}{2}}$.
That is, $\qquad y^2 = 3^{\frac{1}{2}+\frac{1}{2}} = 3$,
and so $\qquad y = \sqrt{3}$.

Hence $3^{\frac{1}{2}}$ ought to mean $\sqrt{3}$, and $x^{\frac{1}{2}}$ ought to mean $\sqrt{x}$.

2. Find the meaning of $5^{\frac{2}{3}}$.

Let $\qquad y = 5^{\frac{2}{3}}$.
Then $\qquad y^2 = y \times y = 5^{\frac{2}{3}} \times 5^{\frac{2}{3}} = 5^{\frac{2}{3}+\frac{2}{3}} = 5^{\frac{4}{3}}$,
and $\qquad y^3 = y^2 \times y = 5^{\frac{4}{3}} \times 5^{\frac{2}{3}} = 5^{\frac{6}{3}} = 5^2$.
Therefore $\qquad y = \sqrt[3]{5^2}$.

Hence $5^{\frac{2}{3}}$ ought to mean $\sqrt[3]{5^2}$, and $x^{\frac{2}{3}}$ ought to mean $\sqrt[3]{x^2}$.

It must not be assumed, without further proof, that $5^{\frac{4}{6}} = 5^{\frac{2}{3}}$, for $5^{\frac{4}{6}}$ means the sixth root of 5^4, that is, the sixth root of 625; while $5^{\frac{2}{3}}$ means the cube root of 5^2, that is, the cube root of 25.

LOGARITHMS

171. Meaning of a Fractional Power. It is now clear that $x^{\frac{p}{q}}$ must mean $\sqrt[q]{x^p}$, that is, the qth root of the pth power of x.

172. Zero Power. The same rule tells us that
$$x^n \cdot x^0 = x^{n+0} = x^n,$$
and hence multiplication by x^0 should leave the value of x unchanged. This shows that *the value of x^0 must be taken as unity, whatever x may be.*

That is, $2^0 = 1$, $(\frac{2}{3})^0 = 1$, $a^0 = 1$, $(-m)^0 = 1$, and so on.

173. Negative Power. Similarly for x^{-n},
$$x^{-n} \cdot x^n = x^{-n+n} = x^0 = 1,$$
and so x^{-n} must be the same as $\dfrac{1}{x^n}$, the *reciprocal* of x^n.

For example, $2^{-3} = \dfrac{1}{2^3} = \dfrac{1}{8}$, and $8^{-\frac{2}{3}} = \dfrac{1}{8^{\frac{2}{3}}} = \dfrac{1}{\sqrt[3]{8^2}} = \dfrac{1}{\sqrt{64}} = \dfrac{1}{4}$.

Exercise 117. Fractional, Zero, and Negative Powers

Using the tables when necessary, calculate the values of:

1. 4^{-2}, 2^{-4}, $27^{\frac{1}{3}}$, $27^{-\frac{1}{3}}$.
2. 10^3, 10^{-3}, $10^{\frac{1}{3}}$, $10^{-\frac{1}{3}}$.
3. 15^2, 15^0, $15^{\frac{1}{2}}$, $15^{-\frac{1}{2}}$.
4. $(-2)^3$, 2^{-3}, $(-2)^{-3}$, $(-2)^0$.

Write each of the following as a power of x:

5. $\sqrt{x}$, $\dfrac{1}{\sqrt{x}}$, $\dfrac{1}{x^2}$, $\dfrac{1}{x^{-2}}$.
6. $\sqrt[3]{x}$, $\sqrt[3]{x^2}$, $\dfrac{1}{\sqrt[3]{x}}$, $\dfrac{1}{\sqrt[3]{x^2}}$.
7. $\sqrt[3]{x^4}$, $\sqrt[3]{\dfrac{1}{x}}$, $\sqrt[3]{\dfrac{1}{x^2}}$.
8. $\sqrt[4]{x}$, $\dfrac{1}{\sqrt[4]{x}}$, $\sqrt[4]{x^5}$, $\dfrac{1}{\sqrt[4]{x^5}}$.

Write each of the following as a root or reciprocal of a root:

9. $x^{\frac{1}{2}}$, $x^{\frac{1}{3}}$, $x^{\frac{1}{4}}$, $x^{-\frac{1}{4}}$.
10. $x^{\frac{3}{5}}$, $x^{\frac{5}{3}}$, $x^{-\frac{3}{5}}$, $x^{-\frac{5}{3}}$.
11. $x^{-\frac{2}{3}}$, $x^{-\frac{1}{2}}$, $x^{-\frac{3}{4}}$.
12. $x^{0\cdot 4}$, $x^{-0\cdot 4}$, $x^{1\cdot 2}$, $x^{-1\cdot 2}$.
13. Write as decimals 10^{-2}, 10^{-3}, 4^{-3}, 4^{-1}, 10^{-4}, 10^{-1}.
14. Write as powers of ten 100, $10{,}000$, $0\cdot 001$, $0\cdot 000001$, $0\cdot 1$.

EXTRACTION OF ROOTS

174. Extraction of Roots. We shall now show that it is possible to extract any root to any degree of accuracy by methods already studied.

Suppose, for example, that we wish to find the fifth root of 589. Raising the integers 1, 2, 3, ... to the fifth power, we find that $3^5 = 243$, and $4^5 = 1024$, so that the required root is between 3 and 4.

Raising 3·1, 3·2, 3·3, ... to the fifth power by long multiplication, we find that $3·5^5 = 525...$, and $3·6^5 = 604...$, so that the required root is between 3·5 and 3·6.

Raising 3·51, 3·52, 3·53, ... to the fifth power, we find that $3·58^5 = 588·0...$, and $3·59^5 = 596·3...$, so that the required root is between 3·58 and 3·59.

Hence the first three figures of $\sqrt[5]{589}$ are 3·58.

In practice there are many devices which shorten the work. It is unnecessary to consider them at present, however, for the student will now be content to accept the tables which follow, knowing from the above example that the main idea of the process by which they can be calculated is within his grasp.

Exercise 118. Extraction of Roots

Find two consecutive integers between which the following lie:

1. $\sqrt{20}$.
2. $\sqrt[3]{40}$.
3. $\sqrt[4]{90}$.
4. $\sqrt[5]{300}$.
5. $\sqrt[6]{900}$.
6. $\sqrt[7]{1200}$.
7. $1500^{\frac{1}{2}}$.
8. $2000^{0·5}$.
9. $50^{0·2}$.
10. $10^{0·8}$.

Find the first two figures of the following:

11. $\sqrt{8}$.
12. $\sqrt[3]{11}$.
13. $\sqrt{0·8}$.
14. $\sqrt[3]{0·8}$.
15. $\sqrt{1·5}$.
16. $\sqrt[3]{1·5}$.
17. $\sqrt[3]{0·25}$.
18. $\sqrt[4]{129}$.
19. $\sqrt[5]{100}$.
20. $\sqrt[7]{225}$.

Find the first three figures of the following:

21. $\sqrt{10}$.
22. $\sqrt[3]{10}$.
23. $\sqrt[4]{10}$.
24. $\sqrt[5]{10}$.
25. $10^{\frac{2}{3}}$.
26. $10^{\frac{3}{4}}$.
27. $10^{\frac{1}{3}}$.
28. $10^{\frac{2}{5}}$.
29. $10^{0·4}$.
30. $10^{0·6}$.

LOGARITHMS

175. Numbers written as Powers of 10. The student will now learn the use of tables which will enable him to write any number as a power of 10, and to find the value of any power of 10.

For example, such tables tell us that, to four significant figures,
$$2 \cdot 316 = 10^{0 \cdot 3647},$$
and that $\quad\quad 10^{1 \cdot 4972} = 31 \cdot 42.$

These tables are of great value in shortening many calculations, particularly those in which multiplications and divisions have to be performed. For example, suppose that we have to multiply $2 \cdot 316$ by $31 \cdot 42$, the two numbers mentioned above. Using the powers of 10 given above we have
$$2 \cdot 316 \times 31 \cdot 42 = 10^{0 \cdot 3647} \times 10^{1 \cdot 4972}$$
$$= 10^{0 \cdot 3647 + 1 \cdot 4972}$$
$$= 10^{1 \cdot 8619}$$

If, now, we have tables which tell us the value of $10^{1 \cdot 8619}$, we can tell the product of $2 \cdot 316$ by $31 \cdot 42$ without actually multiplying, long multiplication being replaced by the much shorter process of addition.

We have such tables, and they tell us that $10^{1 \cdot 8619} = 72 \cdot 77$, so that we know that, to three significant figures at least,
$$2 \cdot 316 \times 31 \cdot 42 = 72 \cdot 77.$$

Their importance having now been shown, we shall consider the actual use of these tables.

176. Logarithm. The power to which 10 must be raised to give any number is called the *logarithm* of that number.

For example, 2 is the logarithm of 100, since $100 = 10^2$;

4 is the logarithm of 10,000, since $10,000 = 10^4$;

-1 is the logarithm of $0 \cdot 1$, since $0 \cdot 1 = \frac{1}{10} = 10^{-1}$;

and $\frac{1}{2}$, or $0 \cdot 5$, is the logarithm of $3 \cdot 162$, since $3 \cdot 162 = \sqrt{10} = 10^{\frac{1}{2}}.$

LOGARITHM TABLES

177. Logarithm Tables. The tables on pages 338-339 give, to four decimal places, the logarithms of all numbers between 1 and 10 having four significant figures. Their use is best shown by some examples.

We should first note that, since $10^0 = 1$ and $10^1 = 10$, the logarithms of all numbers between 1 and 10 lie between 0 and 1.

1. Find the logarithm of 6·38.

Find 6·3 in the left-hand column, and pass along this horizontal line to the column headed 8. Here we find the figures 8048, and these are the figures of the logarithm required. Since the logarithm lies between 0 and 1, it is, therefore, 0·8048; that is, $6·38 = 10^{0·8048}$.

2. Find the logarithm of 6·387.

As in Ex. 1, find the logarithm of 6·38, which is 0·8048. In the "difference columns" at the right-hand side of the page we find, on the same line, the number 5 in the column headed 7. We then proceed thus:

6·38	0·8048
7	5
6·387	0·8053

Therefore the logarithm of 6·387 is 0·8053; that is, $6·387 = 10^{0·8053}$.

3. Find the logarithm of 2·303.

In the line which has 2·3 on the left-hand side, and in the column headed 0, we find 3617. In the same line, and in the difference column headed 3, we find 6. Then, as in Ex. 2,

2·30	0·3617
3	6
2·303	0·3623

Therefore the logarithm of 2·303 is 0·3623; that is, $2·303 = 10^{0·3623}$.

4. Find the logarithm of 9·008.

In the line which has 9·0 on the left-hand side, and in the column headed 0, we find 9542. Proceeding as before, we find $9542 + 4 = 9546$ and hence the required logarithm is 0·9546; that is, $9·008 = 10^{0·9546}$.

LOGARITHMS

Exercise 119. Finding Logarithms

Find the logarithms of the following numbers:

1. 3, 3·2, 3·22, 3·222.
2. 3·02, 3·022, 3·002.
3. 4, 4·1, 4·12, 4·173.
4. 4·01, 4·03, 4·036.
5. 5, 5·0, 5·00, 5·004.
6. 6, 6·7, 6·9, 6·09.
7. 7, 7·111, 7·112, 7·113.
8. 8·0, 8·000, 8·002.
9. 9, 9·007, 9·004, 9·4.
10. 1, 1·7, 1·07, 1·007.
11. 2, 2·2, 2·22, 2·222.
12. 3·7, 3·77, 3·777.
13. 4·5, 4·05, 4·005.
14. 5·3, 5·30, 5·03.
15. 6·1, 6·02, 6·003.
16. 7·3, 7·33, 7·033.
17. 8·8, 8·08, 8·88, 8·088.
18. 9·1, 9·01, 9·111, 9·011.
19. 1·9, 1·09, 1·90, 1·909.
20. 2·04, 2·044, 2·440.
21. 7·65, 7·654, 7·645.
22. 7·605, 7·650, 7·656.
23. 6·0, 6·05, 6·500, 6·505.
24. 6·07, 6·70, 6·700, 6·707.
25. 9·000, 9·00, 9·0, 9·009.
26. 4·7, 4·007, 4·700.
27. 3·00, 3·40, 3·04, 3·046.
28. 1·4, 1·40, 1·04, 1·004.
29. 5·5, 5·005, 5·500.
30. 8·03, 8·30, 8·003, 8·3.
31. 2·3, 2·34, 2·345, 2·045.
32. 6·600, 6·650, 6·654.
33. 1·1, 1·11, 1·111, 1·011.
34. 3·05, 3·053, 3·035.
35. 5·70, 5·07, 5·007, 5·7.
36. 7·89, 7·890, 7·891.
37. 9·876, 9·076, 9·006.
38. 8·364, 9·207, 4·123, 7·7.
39. 1·202, 2·653, 4·179.
40. 5·8, 5·88, 5·888, 5·008.

41. Solve the equation $10^x = 7$.
Since $10^x = 7$,
therefore $x = 0·8451$.

Solve the following equations:

42. $10^x = 3·2$.
43. $10^x = 3·25$.
44. $10^x = 3·257$.
45. $10^{2x-1} = 3·2$.
46. $10^{1-x} = 3·2$.
47. $10^{3x} = 3·2$.
48. $10^x = 4·891$.
49. $10^x = 9·364$.
50. $10^{2x+5} = 6·381$.

FINDING LOGARITHMS

178. Finding the Logarithm of any Number. It is now easy to find the logarithms of numbers which do not lie between 1 and 10. The method is as follows:

1. Find the logarithm of $231{\cdot}7$.

Since
$$231{\cdot}7 = 2{\cdot}317 \times 100$$
$$= 10^{0{\cdot}3649} \times 10^2$$
$$= 10^{2{\cdot}3649},$$

the required logarithm is evidently $2{\cdot}3649$.

2. Find the logarithm of $89{,}700$.

Since
$$89{,}700 = 8{\cdot}97 \times 10{,}000$$
$$= 10^{0{\cdot}9528} \times 10^4$$
$$= 10^{4{\cdot}9528},$$

the required logarithm is $4{\cdot}9528$.

3. Find the logarithm of $0{\cdot}6004$.

Since
$$0{\cdot}6004 = 6{\cdot}004 \times \tfrac{1}{10}$$
$$= 10^{0{\cdot}7785} \times 10^{-1}$$
$$= 10^{-1 + 0{\cdot}7785},$$

the required logarithm is $-1 + 0{\cdot}7785$. It is customary to write this logarithm as $\bar{1}{\cdot}7785$, or as $0{\cdot}7785 - 1$, the subtraction not being performed. Thus the required logarithm is $\bar{1}{\cdot}7785$, or $0{\cdot}7785 - 1$.

4. Find the logarithm of $0{\cdot}00213$.

$0{\cdot}00213 = 2{\cdot}13 \times \tfrac{1}{1000} = 10^{0{\cdot}3284} \times 10^{-3} = 10^{-3 + 0{\cdot}3284} = 10^{\bar{3}{\cdot}3284}.$

Hence the required logarithm is $\bar{3}{\cdot}3284$.

The integral part of the logarithm of any number is the number of places through which the decimal point must be moved in order that it may follow the first significant figure;

This integral part is positive if the decimal point is moved to the left, negative if it is moved to the right;

The decimal part of the logarithm is found from the tables.

The logarithm of x is indicated by the symbol $\log x$.

LOGARITHMS

Exercise 120. Finding Logarithms

Without using the rule, find the logarithms of the following numbers as in the examples on page 277:

1. 25.	6. 64.	11. 425.	16. 1728.
2. 250.	7. 0·64.	12. 42·5.	17. 172·8.
3. 0·25.	8. 640.	13. 0·425.	18. 17·28.
4. 0·025.	9. 0·0064.	14. 0·0425.	19. 0·1728.
5. 25,000.	10. 64,000.	15. 42,500.	20. 172,800.

Using the rule, find the logarithms of the following numbers:

21. 7.	31. 24·7.	41. 41·36.	51. 0·0004.
22. 7·7.	32. 247.	42. 528·4.	52. 0·1004.
23. 7·77.	33. 2470.	43. 5·432.	53. 0·0123.
24. 7·777.	34. 0·247.	44. 68·47.	54. 0·0023.
25. 77·77.	35. 2478.	45. 0·7384.	55. 38,010.
26. 777·7.	36. 247·8.	46. 0·9128.	56. 527·60.
27. 7777.	37. 24·78.	47. 376·8.	57. 0·0007.
28. 0·7777.	38. 2·478.	48. 489·3.	58. 0·00007.
29. 77,770.	39. 0·2478.	49. 0·0009.	59. 0·00203.
30. 777,700.	40. 247,800.	50. 5·007.	60. 0·27070.

Solve the following equations:

61. $10^x = 42$.	65. $10^x = 378·4$.	69. $10^{x+1} = 4273$.
62. $10^x = 42·1$.	66. $10^x = 4976$.	70. $10^{x-1} = 27·34$.
63. $10^x = 42·15$.	67. $10^x = 0·278$.	71. $10^{2x} = 41·32$.
64. $10^x = 421·5$.	68. $10^x = 0·0396$.	72. $10^{2x-1} = 823$.

73. If 10 is raised to the power x^2, the result is 10,000. What is the value of x?

74. Write down the relation of 2, 10, and log 2. Find log 2 and represent this relation by an equation.

ANTILOGARITHMS

179. Antilogarithm. The result of raising 10 to any power n is called the *antilogarithm* of n.

Thus, since $10^2 = 100$, 100 is the antilogarithm of 2;

since $10^{-2} = \dfrac{1}{10^2} = \dfrac{1}{100} = 0\cdot 01$, $0\cdot 01$ is the antilogarithm of -2;

and since $10^{\frac{1}{3}} = \sqrt[3]{10} = 2\cdot 154$, $2\cdot 154$ is the antilogarithm of $\tfrac{1}{3}$.

In other words, the number whose logarithm is n is the antilogarithm of n. Thus

$\log 2 = 0\cdot 3010$, and the antilogarithm of $0\cdot 3010$ is 2.

180. Finding the Antilogarithm of any Number. We find antilogarithms from the tables in the same way as logarithms, as will be seen from the following examples.

1. Find the antilogarithm of $2\cdot 3492$.

Corresponding to $0\cdot 349$ in the table of antilogarithms on page 342 we find 2234, and the number under 2 in the difference column is 1. Hence the figures of the antilogarithm are 2235. Then

$$10^{2\cdot 3492} = 10^2 \times 10^{0\cdot 3492} = 10^2 \times 2\cdot 235 = 223\cdot 5.$$

2. Find the antilogarithm of $1\cdot 3008$.

$$10^{1\cdot 3008} = 10^1 \times 10^{0\cdot 3008} = 10 \times 1\cdot 999 = 19\cdot 99.$$

3. Find the antilogarithm of $\bar{1}\cdot 1101$.

$$10^{\bar{1}\cdot 1101} = 10^{-1} \times 10^{0\cdot 1101} = \tfrac{1}{10} \times 1\cdot 288 = 0\cdot 1288.$$

4. Find the antilogarithm of $\bar{3}\cdot 6131$.

$$10^{\bar{3}\cdot 6131} = 10^{-3} \times 10^{0\cdot 6131} = \tfrac{1}{1000} \times 4\cdot 103 = 0\cdot 004103.$$

From these examples we derive the following rule:

The figures of the antilogarithm of any number are found from the tables, using the decimal part of the number only;

To find the position of the decimal point, begin after the first significant figure of the antilogarithm and move as many places as the integral part of the number, to the right if the integral part is positive, and to the left if it is negative.

LOGARITHMS

Exercise 121. Finding Antilogarithms

Without using the rule, find the antilogarithms of:

1. 1·102.	6. 4·2208.	11. 6·3352.	16. 0·0000.
2. 1·205.	7. 5·3999.	12. $\bar{2}$·5714.	17. 1·0000.
3. 2·114.	8. $\bar{1}$·4508.	13. $\bar{3}$·2136.	18. 10·0000.
4. 2·442.	9. $\bar{3}$·6252.	14. $\bar{1}$·9100.	19. $\bar{1}$·0000.
5. 3·454.	10. 0·7889.	15. 0·0900.	20. $\bar{1}$·1111.

Using the rule, find the antilogarithms of:

21. 0·1200.	31. 6·1000.	41. 4·7780.	51. 0·8162.
22. 0·1230.	32. 6·1200.	42. $\bar{4}$·7782.	52. $\bar{1}$·9236.
23. 0·1234.	33. $\bar{6}$·1200.	43. 2·3648.	53. $\bar{2}$·4173.
24. 1·1234.	34. 0·1257.	44. 1·4763.	54. $\bar{3}$·2992.
25. 5·1234.	35. 3·4462.	45. 0·2980.	55. 4·0020.
26. 2·7130.	36. 2·9865.	46. 0·5236.	56. 2·0130.
27. 2·7132.	37. 3·8726.	47. 4·2998.	57. $\bar{5}$·9000.
28. 3·5240.	38. 4·4027.	48. 3·7261.	58. $\bar{3}$·6235.
29. 3·5245.	39. 2·3333.	49. 1·6111.	59. 4·1209.
30. $\bar{3}$·5245.	40. $\bar{4}$·4330.	50. 3·4444.	60. $\bar{6}$·1350.

Solve the following equations:

61. $10^{0·3010} = x$.	66. $10^{1·8756} = x$.	71. $10^{4·7945} = 2x$.
62. $10^{0·4166} = x$.	67. $10^{2·8370} = x$.	72. $10^{2·7952} = x$.
63. $10^{0·5011} = x$.	68. $10^{3·6693} = x$.	73. $10^{\bar{3}·8624} = x$.
64. $10^{0·8287} = x$.	69. $10^{\bar{1}·9518} = x$.	74. $10^{2·9108} = 2x$.
65. $10^{0·9063} = x$.	70. $10^{\bar{3}·8401} = x$.	75. $10^{3·9276} = 3x$.

Solve the following equations:

76. $\log x = 2·8149$. 78. $\log x = 4·5809$. 80. $\log x = \bar{3}·7750$.
77. $\log x = \bar{1}·8370$. 79. $\log x = 0·6510$. 81. $\log 2x = 1·7871$

MULTIPLICATION

181. Multiplication by Logarithms. Multiplication by logarithms will be understood from the following example:

Find the product of $32 \cdot 5 \times 423 \cdot 1 \times 0 \cdot 0175$.

$$\begin{aligned}
32 \cdot 5 \times 423 \cdot 1 \times 0 \cdot 0175 &= 10^{1 \cdot 5119} \times 10^{2 \cdot 6264} \times 10^{\bar{2} \cdot 2430} \\
&= 10^{1 \cdot 5119 + 2 \cdot 6264 + 0 \cdot 2430 - 2} \\
&= 10^{4 \cdot 3813 - 2} \\
&= 10^{2 \cdot 3813} \\
&= 240 \cdot 6.
\end{aligned}$$

Rough check. $30 \times 400 \times 0.02 = 240$.

Since we do not know the fifth significant figure of the logarithms and antilogarithms, the fourth figure may be slightly in error.

In practice the work would be arranged thus:

$$\begin{aligned}
\log\ 32 \cdot 5 &= 1 \cdot 5119 \\
\log\ 423 \cdot 1 &= 2 \cdot 6264 \\
\log\ 0 \cdot 0175 &= \underline{0 \cdot 2430 - 2} \\
& \ 4 \cdot 3813 - 2 = 2 \cdot 3813 = \log 240 \cdot 6.
\end{aligned}$$

To multiply several numbers together, add their logarithms, and the antilogarithm of the sum gives the required product.

Exercise 122. Multiplication by Logarithms

Multiply the following by use of logarithms, and check:

1. $42 \cdot 3 \times 21 \cdot 7$.
2. $3 \cdot 68 \times 42 \cdot 7$.
3. $51 \cdot 9 \times 3 \cdot 14$.
4. $26 \cdot 7 \times 0 \cdot 319$.
5. $6 \cdot 26 \times 0 \cdot 413$.
6. $52 \cdot 27 \times 2 \cdot 36$.
7. $6 \cdot 133 \times 4 \cdot 48$.
8. $52 \cdot 93 \times 27 \cdot 36$.
9. $284 \cdot 5 \times 1 \cdot 417$.
10. $334 \cdot 6 \times 29 \cdot 41$.
11. $2 \cdot 37 \times 43 \cdot 95$.
12. $576 \cdot 1 \times 293 \cdot 7$.
13. $5 \cdot 916 \times 2 \cdot 789$.
14. $0 \cdot 7392 \times 0 \cdot 6281$.
15. $0 \cdot 09764 \times 287 \cdot 6$.
16. $4 \cdot 31 \times 52 \cdot 73 \times 0 \cdot 97$.
17. $2 \cdot 98 \times 0 \cdot 7623 \times 42 \cdot 8$.
18. $557 \cdot 9 \times 0 \cdot 07234 \times 2 \cdot 9$.
19. $4 \cdot 83 \times 2 \cdot 964 \times 0 \cdot 987$.
20. $5384 \times 2963 \times 51 \cdot 92$.

LOGARITHMS

182. Division by Logarithms. Division by logarithms will be understood from the following examples:

1. Required to divide 67·39 by 1·994.

$$\frac{67\cdot 39}{1\cdot 994} = \frac{10^{1\cdot 8286}}{10^{0\cdot 2998}} = 10^{1\cdot 8286 - 0\cdot 2998} = 10^{1\cdot 5288} = 33\cdot 79.$$

In practice the work would be arranged thus:

$$\log 67\cdot 39 = 1\cdot 8286$$
$$\log 1\cdot 994 = 0\cdot 2998$$
$$\overline{1\cdot 5288} = \log 33\cdot 79.$$

Hence $67\cdot 39 \div 1\cdot 994 = 33\cdot 79$, to four significant figures.

2. Required to divide 67·39 by 1·994, using reciprocals.

$$\log \frac{1}{1\cdot 994} = \log \frac{0\cdot 1}{0\cdot 1994} = -1 + \log \frac{1}{0\cdot 1994} = -1 + \log \text{recip } 0\cdot 1994.$$

From page 340, log recip 0·1994 = 0·7002.

$$\log \quad 67\cdot 39 = 1\cdot 8286$$
$$\log \text{recip } 1\cdot 994 = \underline{0\cdot 7002 - 1}$$
$$1\cdot 5288 \quad = \log 33\cdot 79.$$

Hence $67\cdot 39 \div 1\cdot 994 = 33\cdot 79$, to four significant figures.

To divide one number by another, subtract the logarithm of the divisor from the logarithm of the dividend, and the antilogarithm of the difference gives the required quotient.

Exercise 123. Division by Logarithms

Divide the following by use of logarithms, and check:

1. $425 \div 2\cdot 5$.
2. $425\cdot 7 \div 2\cdot 53$.
3. $42\cdot 57 \div 25\cdot 3$.
4. $4\cdot 257 \div 0\cdot 253$.
5. $2963 \div 5125$.
6. $2\cdot 963 \div 0\cdot 05125$.
7. $3849 \div 1962$.
8. $38\cdot 49 \div 1\cdot 962$.
9. $0\cdot 3849 \div 0\cdot 01962$.
10. $0\cdot 03849 \div 19\cdot 62$.
11. $5\cdot 28 \div 62\cdot 7$.
12. $39\cdot 72 \div 41\cdot 68$.
13. $4\cdot 783 \div 5\cdot 296$.
14. $3\cdot 772 \div 0\cdot 299$.
15. $52\cdot 87 \div 0\cdot 0037$.
16. $0\cdot 0762 \div 4\cdot 003$.
17. $0\cdot 09111 \div 0\cdot 1724$.
18. $6\cdot 368 \div 52\cdot 93$.

RAISING TO A POWER

183. Raising to a Power by Logarithms. Any power of a number is easily found by logarithms.

1. Raise $4 \cdot 63$ to the seventh power.

Since $\log 4 \cdot 63 = 0 \cdot 6656$, we have
$$4 \cdot 63^7 = (10^{0 \cdot 6656})^7 = 10^{0 \cdot 6656 \times 7}$$
$$= 10^{4 \cdot 6592}.$$

Since the antilogarithm of $4 \cdot 6592$ is $45{,}620$, we know that
$$4 \cdot 63^7 = 45{,}620, \text{ to three significant figures.}$$

2. Find the value of $0 \cdot 02946^4$.

As in Ex. 1, $\log 0 \cdot 02946 = 0 \cdot 4692 - 2$
$$\frac{4}{1 \cdot 8768 - 8} = \overline{7} \cdot 8768.$$

The antilogarithm of $\overline{7} \cdot 8768$ is $0 \cdot 0000007530$.
Therefore $0 \cdot 02946^4 = 0 \cdot 0000007530$, to three significant figures.

3. Find the value of $12 \cdot 7^2 \times 34 \cdot 61^3$.

$\log 12 \cdot 7 \ = 1 \cdot 1038$ $\log 12 \cdot 7^2 \ = 2 \log 12 \cdot 7 \ = 2 \cdot 2076$
$\log 34 \cdot 61 = 1 \cdot 5392$ $\log 34 \cdot 61^3 = 3 \log 34 \cdot 61 = 4 \cdot 6176$
$$\overline{6 \cdot 8252}$$
Antilogarithm of $6 \cdot 8252 = 6{,}687{,}000$.
Therefore $12 \cdot 7^2 \times 34 \cdot 61^3 = 6{,}687{,}000$, to three significant figures.

To find a power of a number, multiply the logarithm of the number by the index of the power, and the antilogarithm of the product gives the required power.

Exercise 124. Finding Powers by Logarithms

Perform the operations indicated:

1. $5 \cdot 2^6$.
2. $5 \cdot 23^4$.
3. $5 \cdot 237^5$.
4. $523 \cdot 7^5$.
5. $0 \cdot 05237^5$.
6. $62 \cdot 39^3$.
7. $6 \cdot 239^3$.
8. $0 \cdot 006239^3$.
9. $43 \cdot 72^6$.
10. $0 \cdot 4372^6$.
11. $3 \cdot 42^3 \times 2 \cdot 98^2$.
12. $5 \cdot 346^4 \times 2 \cdot 936^3$.
13. $28 \cdot 47^3 \div 39 \cdot 52^4$.
14. $3 \cdot 463^4 \div 2 \cdot 841^3$.
15. $2 \cdot 776^3 \times 0 \cdot 294^2$.

LOGARITHMS

184. Finding a Root by Logarithms. Any root of a number is easily found by logarithms.

1. Find the fifth root of $396 \cdot 7$.

Since $\log 396 \cdot 7 = 2 \cdot 5985$, we have
$$\sqrt[5]{396 \cdot 7} = \sqrt[5]{10^{2 \cdot 5985}}.$$

By § 170, $\quad \sqrt[5]{10^{2 \cdot 5985}} = 10^{\frac{2 \cdot 5985}{5}} = 10^{0 \cdot 5197}.$

Hence $0 \cdot 5197$ is the logarithm of $\sqrt[5]{396 \cdot 7}$, or
$$\sqrt[5]{396 \cdot 7} = \text{antilogarithm of } 0 \cdot 5197 = 3 \cdot 309.$$

2. Find the seventh root of $0 \cdot 0143$.

From the tables, $\log 0 \cdot 0143 = 0 \cdot 1553 - 2$.

It is not convenient to divide this by 7, on account of the negative integral part 2. But since $-2 = 5 - 7$, we write
$$\log 0 \cdot 0143 = 5 \cdot 1553 - 7.$$
$$\tfrac{1}{7} \log 0 \cdot 0143 = 0 \cdot 7365 - 1 = \log 0 \cdot 5451.$$
Therefore $\quad \sqrt[7]{0 \cdot 0143} = 0 \cdot 5451.$

To find the nth root of a number, divide the logarithm of the number by n, and the antilogarithm of the result gives the required root.

Exercise 125. Finding Roots by Logarithms

Find the following roots by use of logarithms:

1. $\sqrt{1 \cdot 923}.$
2. $\sqrt{0 \cdot 2746}.$
3. $\sqrt{29 \cdot 37}.$
4. $\sqrt[3]{41 \cdot 96}.$
5. $\sqrt[3]{0 \cdot 0734}.$
6. $\sqrt[4]{2 \cdot 911}.$
7. $\sqrt[4]{0 \cdot 6672}.$
8. $\sqrt[5]{647 \cdot 3}.$
9. $\sqrt{48 \cdot 34}.$
10. $\sqrt[5]{1 \cdot 963}.$
11. $\sqrt[5]{0 \cdot 8274}.$
12. $\sqrt[5]{0 \cdot 0436}.$
13. $\sqrt[5]{0 \cdot 0074}.$
14. $\sqrt[5]{0 \cdot 0006}.$
15. $\sqrt[6]{28 \cdot 4}.$
16. $\sqrt[6]{3 \cdot 42}.$
17. $\sqrt[6]{0 \cdot 48}.$
18. $\sqrt[6]{0 \cdot 006}.$
19. $\sqrt[7]{12420}.$
20. $\sqrt[7]{34 \cdot 63}.$
21. $\sqrt[7]{8 \cdot 112}.$
22. $\sqrt[8]{2}.$
23. $\sqrt[8]{3}.$
24. $\sqrt[8]{0 \cdot 074}.$
25. $\sqrt[9]{64 \cdot 36}.$
26. $\sqrt[9]{7 \cdot 438}.$
27. $\sqrt[10]{82 \cdot 6}.$
28. $\sqrt[25]{4865}.$

USE OF LOGARITHMS

185. Use of Logarithms. The following examples show how logarithms may be applied:

1. Find the volume of a sphere whose diameter is 7·3 in.

From § 14, $v = \frac{4}{3}\pi r^3$. Since $r = \frac{1}{2}d$, this reduces to $\frac{1}{6}\pi d^3$.
We have, then, to find the value of $\frac{1}{6}\pi .7\cdot 3^3$.
From the table on page 344, $\log \pi = 0\cdot 4971$.
We also know that $\frac{1}{6} = 0\cdot 1667$.

$$\log 0\cdot 1667 = 0\cdot 2219 - 1$$
$$\log \pi = 0\cdot 4971$$
$$3 \log 7\cdot 3 = 2\cdot 5899$$
$$\overline{3\cdot 3089 - 1} = 2\cdot 3089 = \log 203\cdot 7.$$

Therefore the required volume is 203·7 cu. in.

2. Find the edge of a cube which has the same volume as a sphere of radius 15 in.

Since $v = \frac{4}{3}\pi r^3 = \frac{4}{3}\pi 15^3$, the edge is $\sqrt[3]{\frac{4}{3}\pi 15^3}$. We then have

$$\log 4 = 0\cdot 6021$$
$$\log \text{recip } 3 = 0\cdot 5229 - 1$$
$$\log \pi = 0\cdot 4971$$
$$3 \log 15 = 3\cdot 5283$$
$$\overline{\log v = 4\cdot 1504}$$
$$\log v = 1\cdot 3835 = \log 24\cdot 18.$$

Therefore the edge of the cube is 24·18 in.

Exercise 126. Applications

1. Find the surface of a sphere of radius 4·7 in.

2. Find the volume of a sphere of radius 7·2 in.

3. Assuming the earth to be a sphere of radius 3956 ml., find the surface and the volume to four significant figures.

4. In the equation $l = ar^{n-1}$, find the value of l when $a = 1\cdot 7$, $r = 2\cdot 8$, and $n = 12$.

5. In the equation $r = \sqrt[n-1]{\dfrac{l}{a}}$, find the value of r when $l = 92\cdot 8$, $a = 3\cdot 7$, and $n = 9$.

LOGARITHMS

6. In the equation $s = \frac{1}{2}gt^2$, find the value of s if $g = 32$ and $t = 42 \cdot 7$.

7. Find the capacity in cubic inches of a pipe $48 \cdot 1$ in. long which has an internal diameter of $3 \cdot 4$ in.

8. If steel wire will support a weight of 215,000 lb. per square inch of cross-section, how many pounds can be lifted by a steel wire of diameter $\frac{7}{48}$ in.?

9. The weight of a cube of a certain kind of stone of edge x feet is $1 \cdot 5 \, x^3$ hundredweight. What is the weight of a cube of the stone of edge $0 \cdot 75$ ft.?

10. What is the area of the cross-section of a steel shaft of diameter $3 \cdot 8$ in.?

11. What is the weight of a steel cylinder 6 ft. long and $2 \cdot 6$ in. in diameter, steel weighing $0 \cdot 28$ lb. per cubic inch?

12. In finding the horse-power of a certain pump it becomes necessary to find the value of $4 \cdot 2 \, G^3/d^4$, where $G = 375$ and $d = 4 \cdot 2$. Find this value.

13. The horse-power of a certain engine is $v \, W/55$, where $v = 48 \cdot 7$, and $W = 224 \cdot 7$. Find the horse-power.

14. The weight in hundredweight which can be carried on a certain beam is $70 \, t^3/L$, where t is the thickness in inches and L is the length in feet. Find the weight if $t = 8 \cdot 2$ and $L = 16 \cdot 5$.

15. A pendulum l inches long takes $0 \cdot 16 \sqrt{l}$ seconds to pass from one end of its swing to the other. What is the number of seconds if the pendulum is $48 \cdot 7$ in. long?

16. A body falls $16 \, t^2$ feet in t seconds. How many feet will it fall in $37 \cdot 7$ sec.?

17. A cube of iron whose edge is $2 \cdot 7$ in. is melted and moulded into a sphere. Find the surface and the volume of the sphere.

18. A sphere of iron whose radius is $4 \cdot 8$ in. is melted and moulded into a cube. Find the surface of the cube.

Printed by BoD™in Norderstedt, Germany